Stoke-on-Trent Libraries
Approved for Sale

HORACE BARKS
REFERENCE LIBRARY

STOKE-ON-TRENT

CERAMIC MATERIALS FOR ADVANCED HEAT ENGINES

CERAMIC MATERIALS FOR ADVANCED HEAT ENGINES

Technical and Economic Evaluation

by

D.C. Larsen, J.W. Adams
IIT Research Institute
Chicago, Illinois

L.R. Johnson, A.P.S. Teotia, L.G. Hill
Argonne National Laboratory
Argonne, Illinois

NOYES PUBLICATIONS
Park Ridge, New Jersey, U.S.A.

Sole distribution by:
Gothard House Publications
Gothard House, Henley-on-Thames,
Oxon RG9 1AJ
Tel: 0491 573602

Copyright © 1985 by Noyes Publications
Library of Congress Catalog Card Number 85-4900
ISBN: 0-8155-1029-2
Printed in the United States

Published in the United States of America by
Noyes Publications
Mill Road, Park Ridge, New Jersey 07656

10 9 8 7 6 5 4 3 2 1

Library of Congress Cataloging in Publication Data
Main entry under title:

Ceramic materials for advanced heat engines.

   Bibliography: p.
   Includes index.
   1. Ceramic materials. 2. Heat-engines--Materials.
I. Larsen, David C.
TJ225.5.C47 1985   621.402'5   85-4900
ISBN 0-8155-1029-2

Foreword

Ceramic materials for advanced heat engines are described in this technological and economic evaluation. Advanced power systems for vehicles can potentially produce fuel efficiencies greatly exceeding those of today's gasoline and diesel engines. Current heat engine technologies, however, are limited by problems involving mechanical strength at high temperatures. Structural ceramics, if they can be reliably mass-produced, can make possible improved vehicle fuel efficiencies through higher-temperature operation and reduced vehicle weight.

In Part I of the book state-of-the-art technical ceramics, mainly silicon carbide and silicon nitride, that have potential as structural components in advanced heat engines, are evaluated. Thermal and mechanical property data were generated on candidate materials, and the results were interpreted with respect to microstructure, purity, secondary phases, environmental effects, and processing methods. Properties measured include: flexural strength, elastic modulus, stress-strain, fracture toughness, creep, oxidation, thermal expansion, thermal diffusivity, thermal shock, and stress rupture.

In Part II the macroeconomic impacts of structural ceramics are modeled for two scenarios, one in which the U.S. dominates the commercialization of ceramics in heat engines throughout the 1990s, and the other in which Japan dominates.

The information in the book is from:

Property Screening and Evaluation of Ceramic Turbine Materials by D.C. Larsen and J.W. Adams of IIT Research Institute for Air Force Wright Aeronautical Laboratories, Materials Laboratory, Air Force Systems Command, April 1984.

A Structural Ceramic Research Program: A Preliminary Economic Analysis by L.R. Johnson, A.P.S. Teotia and L.G. Hill of Energy and Environmental Systems Division, Center for Transportation Research, Argonne National Laboratory for the U.S. Department of Energy, March 1983.

The table of contents is organized in such a way as to serve as a subject index and provides easy access to the information contained in the book.

Advanced composition and production methods developed by Noyes Publications are employed to bring this durably bound book to you in a minimum of time. Special techniques are used to close the gap between "manuscript" and "completed book." In order to keep the price of the book to a reasonable level, it has been partially reproduced by photo-offset directly from the original reports, and the cost saving passed on to the reader. Due to this method of publishing, certain portions of the book may be less legible than desired.

NOTICE

The materials in this book were prepared as accounts of work sponsored by the U.S. Department of Energy and the Air Force Systems Command. Publication does not signify that the contents necessarily reflect the views and policies of the contracting agencies or the publisher, nor does mention of trade names or commercial products constitute endorsement or recommendation for use.

Where Government drawings, specifications, or other data are used for any purpose other than in connection with a definitely related Government procurement operation, the United States Government thereby incurs no responsibility nor any obligation whatsoever; and the fact that the government may have formulated, furnished, or in any way supplied the said drawings, specifications, or other data, is not to be regarded by implication or otherwise as in any manner licensing the holder or any other person or corporation, or conveying any rights or permission to manufacture, use, or sell any patented invention that may in any way be related thereto.

Contents and Subject Index

PART I
MATERIALS EVALUATION

ACKNOWLEDGMENTS....................................2

1. INTRODUCTION......................................4

2. MATERIALS...8

3. TEST PLAN..18

4. MATERIALS CHARACTERIZATION.......................21

5. TEST METHODOLOGY.................................30
   Reflected Light Microscopy........................30
   Flexural Strength and Elastic Modulus.............32
   Creep and Stress Rupture..........................39
   Dynamic Elastic Moduli............................39
   Thermal Shock/Internal Friction...................43
   Thermal Expansion.................................44
   Thermal Diffusivity...............................46

6. MICROSTRUCTURE, ROOM-TEMPERATURE STRENGTH, AND
   ELASTIC PROPERTIES................................51
   Si_3N_4 Materials...............................51
      Microstructural Features.......................51
      Flexural Strength/Fracture Sources and Elastic Properties........66
   SiC Materials.....................................74
      Strength-Grain Size Relation...................78

Effect of Porosity 99
Fracture Origins 100

7. ELEVATED TEMPERATURE STRENGTH AND TIME DEPENDENCE 101
Silicon Nitride Materials 103
    Hot-Pressed Si_3N_4 103
    Reaction-Sintered Si_3N_4 140
    Sintered Si_3N_4 146
SiC Materials 146
Summary of High Temperature Fracture Origins in Si_3N_4 and SiC ... 164

8. CREEP BEHAVIOR 169
HP-Si_3N_4 170
RS-Si_3N_4 Materials 179
SiC Materials 179

9. THERMAL EXPANSION 181

10. THERMAL SHOCK RESISTANCE 187

11. OXIDATION BEHAVIOR 200
Si_3N_4 212
    MgO-Doped HP-Si_3N_4 212
    CeO_2-Doped HP-Si_3N_4 215
    Y_2O_3-Doped HP-Si_3N_4 218
    RS-Si_3N_4 Materials 225
SiC Materials 234
    HP-SiC 238
    Sintered SiC 241
    Silicon-Infiltrated SiC 245
Summary of Oxidation in Silicon Ceramics 248

12. ZIRCONIA CERAMICS FOR DIESEL ENGINES 249
The Advantages of Zirconia in Diesel Engines 249
Polymorphism and Phase Stabilization in Zirconia .. 250
Concept of Transformation-Toughening 251
Evaluation of Various Commercial Zirconia Materials 253
    Density, Phase Assemblage 253
    Microstructure 256
    Strength, Deformation, and Fracture 256
    Thermal Expansion, Long-Term Stability 267
    Conclusion 279

13. CONCLUDING REMARKS 280
Silicon-Base Ceramics 280
Transformation-Toughened Zirconia 282

Contents and Subject Index ix

REFERENCES 283

APPENDIX: REFLECTED LIGHT AND SEM MICROGRAPHS OF
POLISHED REACTION-SINTERED SILICON NITRIDE MATERIALS ... 293

PART II
ECONOMIC EVALUATION

ACKNOWLEDGMENTS 314

1. INTRODUCTION 315

2. ECONOMIC IMPACTS OF STRUCTURAL CERAMIC
   APPLICATIONS 319
   Methodology Used to Assess Impacts 319
   Potential Uses for High-Temperature Structural Ceramics ... 320
   Market Penetration for Selected Applications 322
      Selection of the Applications for Analysis 322
         Heat Engine Ceramic Components 322
         Ceramic Bearings 322
         Turbochargers 323
         Gas Turbines 323
      Market Potential for Structural Ceramic Applications ... 323
      Estimated Market Penetration of Structural Ceramic
         Applications 323
   **Economic Effects of Alternative Scenarios for Ceramic
   Commercialization** 325
      Assumptions 327
      Macroeconomic Effects of Ceramic Penetration in the Heat
         Engine Market 329
         Base Case Macroeconomic Scenario 329
         Macroeconomic Effects—U.S. Dominant vs Foreign Dominance
            Cases 330
               Real Gross National Product 330
               Total U.S. Employment 331
               Total U.S. Imports Savings 332
               Total Fuel Savings in the Economy 332
               Import Fuel-Bill Savings 333
               Conclusions 334
   Strategic Materials Impacts 335
      Beryllium 335
      Cobalt 339
      Chromium 339
      Manganese 341
      Nickel 341
      Columbium (Niobium) 341
      Platinum and Palladium 342
      Tantalum 342

Titanium ... 342
Tungsten ... 343
Summary of Overall Impacts ... 343
Foreign Competition ... 343

3. THE NEED FOR FEDERAL RESEARCH SUPPORT ... 346
Assessment of Private Sector Support ... 346
Research and Development: Investment vs "Borrowing" ... 348
Productivity and Basic Research ... 348
The Relationship Between Basic Research, Innovation, and Diffusion ... 349
A Question of "Borrowing" ... 350
Ceramics: A Preliminary Look ... 350

4. INDUSTRY PERSPECTIVES ... 353
The Market Potential for Structural Ceramics ... 353
The Need for Research and Development ... 356
The Threat of Foreign Competition ... 359

5. CONCLUSIONS ... 362

APPENDIX A: CHANGES IN THE ECONOMIC MODEL ... 365
The DRI Annual Model of the U.S. Economy ... 365
Changes Made in DRI Base Case Scenario for the U.S. Dominant Case ... 365
Changes Made in DRI Base Case Scenario for the Foreign Dominance Case ... 367
Changes in the Projections of the U.S. Economy ... 368

APPENDIX B: TRANSLATIONS OF JAPANESE NEWSPAPER ARTICLES ... 370
Toyota Utilizes Ceramic Metal Fiber Composite for Motor Parts ... 370
The Battlefront of Advanced Technology Development ... 371

APPENDIX C: SYNOPSIS OF AMERICAN COUNTERATTACK—PLAN Z, 1985 ... 374

REFERENCES ... 378

Part I

Materials Evaluation

The information in Part I is from *Property Screening and Evaluation of Ceramic Turbine Materials* by D.C. Larsen and J.W. Adams of IIT Research Institute for Air Force Wright Aeronautical Laboratories, Materials Laboratory, Air Force Systems Command, April 1984.

Acknowledgments

IITRI personnel who have contributed to this program include J. L. Sievert, B. N. Norikane, G. T. Jeka, W. R. Logan, H. H. Nakamura, and J. Berlanga. The many discussions on the subject of the properties and microstructure of technical ceramics with S. A. Bortz, Y. Harada, and R. F. Firestone, are acknowledged.

Thanks are due the organizations that have contributed test samples to this program. The various personnel in the organizations that have contributed more than sixty materials to this program are listed in the accompanying table. Their contributions and cooperation are greatly appreciated.

We are indebted to Dr. H. Priest and Mr. F. Burns at USAMMRC for supplying oxygen content data. Also, thanks to Mr. J. Muntz of AFWAL for supplying cation impurity analyses performed by spectrographic and atomic absorption techniques. Additionally, the continued support and aid of Mr. D. Zabierek and Capt. P. Coty of AFAPL, and Mr. N. M. Geyer, Dr. K. S. Mazdiyasni, and Dr. P. Land of AFWAL is acknowledged. IITRI is also grateful for the added support from Dr. S. Wax of AFOSR. The direction provided by Dr. H. C. Graham of AFWAL also deserves mention.

Finally, special appreciation for the many helpful discussions and active collaboration with the Air Force Project Manager, Dr. Robert Ruh of the Processing and High Temperature Materials Branch of AFWAL's Metals and Ceramics Division. Dr. Ruh served both as project monitor as well as co-investigator on many aspects of this program.

PRINCIPAL VENDOR CONTACTS

Organization	Contacts
AiResearch	K. H. Styhr
Annawerk	E. Gugel
Associated Engineering Development	J. C. Moore
Battelle	R. R. Wills
Carborundum	G. W. Weber
Ceradyne	J. Rubin
Chemetal/SFL	J. J. Stiglich
Coors	D. Roy, K. Green, B. Seegmiller
Cummins Engine	R. Kamo
ESK	K. A. Schwetz
Fiber Materials	P. F. Jahn
Ford	A. F. McLean
General Electric	C. A. Johnson, W. A. Nelson
Georgia Tech	J. N. Harris
GTE	C. L. Quackenbush
Harbison-Walker	D. Petrak
Indussa/Nippon Denko	N. Ramesh
Kawecki-Berylco	R. J. Longenecker
Kyocera	W. Everitt
NASA-Lewis	S. N. Dutta
Naval Research Laboratory	R. W. Rice
Norton	R. Bart, M. L. Torti
Raytheon	J. S. Waugh
Rocketdyne	G. Schnittgrund
Toshiba	T. Ochiai
UKAEA	W. A. McLaughlan
U.S. Bureau of Mines	H. Heystek
Westinghouse	R. Bratton

1. Introduction

Silicon-base ceramics, silicon nitride and silicon carbide, and zirconia ceramics have potential structural application in advanced heat engines such as gas turbines and diesel engines. Components include rotor blades, stator vanes, combustion chambers, piston caps, and cylinder liners. Such utilization of ceramics offers several potential advantages, including higher temperature operation leading to increased thermal efficiency and decreased specific fuel consumption, decreased weight and thus lower stresses in rotating components, greater thrust-to-weight ratio, lower potential life cycle cost, decreased complexity through the use of noncooled components, and reduced dependency on the use of strategic materials (e.g., cobalt and chromium) that are used in metallic superalloys.

Over the past decade the various forms of Si_3N_4 and SiC have been the most promising materials for advanced heat engine application. These materials are high strength, oxidation-resistant, and thermal shock resistant. Improved versions of SiC and Si_3N_4 are currently being developed within the ceramics industry. These materials have enjoyed some success, although limited, as structural components in the hot section of prototype and demonstration engines. For ceramics to become a reality in the intended application, various requirements must be met including: the development of overall life prediction methodology; employment of realistic design methodology; and demonstration of fabrication process feasibility for the required ceramic component configurations. What is currently lacking for structural ceramics is uniformity, reproducibility, and reliability. Much work is being performed to develop an overall life prediction methodology for various ceramic component designs. The biggest challenges facing designers in the use of structural ceramics include: the statistical strength distribution; low fracture toughness and surface sensitivity; the existence of time-dependent effects such as subcritical crack growth and associated strength degradation; the lack of data regarding the combined environmental effects of oxidation, corrosion, erosion, and deposition from the fuel combustion products; batch-to-batch variability resulting from inadequate process control; and lack of an NDE technology capable of detecting critical strength-limiting flaws.

These challenges are compounded by the fact that new candidate materials for advanced heat engine applications are continually emerging, and their properties are strongly dependent on microstructure, purity, and processing history. Therefore, this program was established with the purpose of supporting various current component design efforts through the comprehensive screening characterization of a wide variety of silicon-base ceramics that have potential use as components in high-temperature gas turbines, radome structures, and other high performance/severe environment applications. There is clearly a need to continue property generation and microstructural studies of these materials that are in a continual stage of improvement.

In this program, thermal and mechanical property data were measured on candidate materials, and the results interpreted with respect to microstructure, purity, secondary phases, environmental effects, and processing methods. Measured properties were related to microstructure. Failure modes and the nature of the critical strength-controlling flaws were determined by optical and SEM fracture surface analysis. Structure-property relations were developed, and generated data were compared to literature data. To date, over sixty materials have been extensively characterized. Properties measured include: flexure strength, elastic modulus, stress-strain, fracture toughness, creep, oxidation, thermal expansion, thermal diffusivity, thermal shock, and stress rupture. Emphasis is placed on defining the limiting aspects of each material, such as the time-dependent strength relating to the existence of operable subcritical crack growth mechanisms. Property measurements were made up to 1500°C in air atmosphere.

This program has the same objective and scope as its predecessor, AFML-TR-79-4188.[1] This report summarizes and overviews the pertinent properties obtained on over sixty (60) silicon-base and zirconia ceramics that are candidates for advanced heat engine applications. Emphasis is placed on predominant behavioral trends for each material type to aid in materials selection. Details of this work may be found in the widely distributed semiannual interim technical reports issued on this program.[2-6] Additionally, various papers have been presented and published that deal with specific aspects of this program.[7-18]

It is noted that zirconia materials were included on this program. These materials in transformation-toughened forms are currently receiving much attention for potential use in diesel engines.

This program involved much more SEM microscopy and microstructural analysis than its predecessor. Reflected light microstructural analysis reveals grain size and porosity distribution. These aspects of microstructure are directly correlated with the strength of ceramics. We cannot overemphasize the importance of SEM examination of fracture surfaces to reveal fracture mode and fracture origins.

This properties screening and evaluation work is a continuing effort in this laboratory. A follow-on program with similar objectives and scope has been initiated.

In the continuing work, the major emphasis is on ceramic-matrix composites, however.

This report is organized as follows. The first two sections that follow list the specific materials studied on this program and the test plan. The chemical composition and phase assemblage of the materials are then presented in Section 4. Section 5 contains the details of the test methodology used on this program. Photographs of specific test equipment are included. Section 6 presents the room-temperature strength and elastic properties of the SiC and Si_3N_4 materials, as correlated to the nature of their microstructure (grain size and porosity). Typical fracture origins for each material are discussed. Elevated temperature strength behavior is presented in Section 7. Here we correlate the high temperature strength and deformation behavior with the fracture mode and the existence of oxide intergranular phases, leading to subcritical crack growth or other forms of static fatigue. Sections 8-11 contain information on (1) creep, (2) thermal expansion, (3) thermal shock, and (4) oxidation behavior. Finally, Section 12 presents our evaluation of various foreign and domestic technical zirconia ceramics.

2. Materials

A wide variety of silicon-base ceramics have emerged during recent years that have potential application as components in high temperature gas turbines and other advanced heat engines. Silicon nitride and silicon carbide, in hot-pressed, reaction bonded, and pressureless sintered forms are currently the most promising materials for this application. These materials range from commercially available (which in most cases are in a continual state of development) to highly developmental. Also, certain materials have been specifically developed for lower cost, less stringent applications. Thus, the intended application and the processing maturity must be considered when comparing materials performance.

Available Si_3N_4 and SiC bodies represent a large family of materials with wide property variation and broad response characteristics during interaction with severe environments. Even though some are termed commercially available, they are in a continual state of development and change. In general, property differences between materials of a given type are directly related to: (1) existence of secondary phases, (2) purity, (3) microstructural aspects such as pore size and distribution, grain size, etc., (4) phase stability, (5) microstructural stability, (6) existence of subcritical crack growth, and (7) the nature of the critical strength-controlling flaws (e.g., contaminant inclusions, pores/pore agglomerates, large grains, unreacted particles, impurity particles, etc.) and how the critical flaws change with time, temperature, and environmental exposure.

Table 1 contains a listing of all materials evaluated on this program (and its predecessor) arranged according to material type (excluding various developmental AFWAL materials).

All materials, with the exception of the Norton materials

investigated earlier (Norton NC-132, NC-435, and NC-350*), were supplied without charge by the respective manufacturers.

In cases where material billets rather than assintered or machined test samples were supplied, AFWAL arranged for machining under a separate contract.

Table 2 contains density and machining information on all materials, as well as the date each material was received. The date received is important since several materials are termed commercially available but are actually still being developed. Some materials have undergone various processing modifications recently which have resulted in improved properties. The indication of the date received thus permits an assessment of the performance of the material as determined in this program, with respect to its maturity level. Some suppliers have made significant improvements in processing technology over the duration of this program, and have supplied new upgraded versions of their materials. It is expected that this will continue with various manufacturers as improved processing techniques yield upgraded materials.

---

*Early on the previous program, the work scope allowed for test samples to be purchased. On the present program, however, funding was not available for this.

TABLE 1. MATERIALS EVALUATED TO DATE

A. **Hot-Pressed Si_3N_4**

- Norton NC-132 HP-Si_3N_4 (1% MgO)
- Norton NCX-34 HP-Si_3N_4 (8% Y_2O_3)
- Harbison-Walker HP-Si_3N_4 (10% CeO_2)
- Kyocera SN-3 HP-Si_3N_4 (4% MgO, 5% Al_2O_3)
- Naval Research Laboratory HP-Si_3N_4 (4-16 wt% ZrO_2)
- AFML Developmental HP-Si_3N_4 (CeO_2 and BN Additives)
- Ceradyne Ceralloy 147A, HP-Si_3N_4 (1% MgO)
- Ceradyne Ceralloy 147Y, HP-Si_3N_4 (15% Y_2O_3)
- Ceradyne Ceralloy 147Y-1, HP-Si_3N_4 (8% Y_2O_3)
- Fiber Materials HP-Si_3N_4 (4% MgO)
- Toshiba HP-Si_3N_4 (4% Y_2O_3, 3% Al_2O_3)
- Toshiba HP-Si_3N_4 (3% Y_2O_3, 4% Al_2O_3, SiO_2)
- Westinghouse HP-Si_3N_4 (4% Y_2O_3, SiO_2)
- NASA/AVCO/Norton HP-Si_3N_4 (10% ZrO_2)

B. **Hot Isostatic Pressed Si_3N_4**

- Battelle HIP-Si_3N_4 (5% Y_2O_3)

C. **Sintered Si_3N_4**

- Kyocera SN-205 Sintered Si_3N_4 (5% MgO, 9% Al_2O_3)
- Kyocera SN-201 Sintered Si_3N_4 (4% MgO, 7% Al_2O_3)
- GTE Sintered Si_3N_4 (6% Y_2O_3)
- AiResearch Sintered Si_3N_4 (8% Y_2O_3, 4% Al_2O_3)
- Rocketdyne SN-50 Sintered Si_3N_4 (6% Y_2O_3, 4% Al_2O_3)
- Rocketdyne SN-104 and SN-46 Sintered Si_3N_4 (14% Y_2O_3, 7% SiO_2)

D. **Reaction Sintered Si_3N_4**

- 1976 Norton NC-350 RS-Si_3N_4
- Kawecki-Berylco RS-Si_3N_4

TABLE 1 (cont.)

D. Reaction Sintered Si_3N_4 (cont.)
- Ford Injection Molded RS-Si_3N_4
- AiResearch Slip Cast RS-Si_3N_4 (Airceram RBN-101)
- Raytheon Isopressed RS-Si_3N_4
- Indussa/Nippon Denko RS-Si_3N_4
- AiResearch Injection Molded RS-Si_3N_4 (Airceram RBN-122)
- 1979 Norton NC-350 RS-Si_3N_4
- Annawerk Ceranox NR-115H RS-Si_3N_4
- Associated Engineering Developments (AED) Nitrasil RS-Si_3N_4
- Georgia Tech RS-Si_3N_4
- AME RS-Si_3N_4
- AiResearch RBN-104 RS-Si_3N_4

E. Hot-Pressed SiC
- Norton NC-203 HP-SiC (~2% Al_2O_3)
- Ceradyne Ceralloy 146A, HP-SiC (2% Al_2O_3)
- Ceradyne Ceralloy 146I, HP-SiC (2% B_4C)

F. Sintered SiC
- General Electric Sintered β-SiC
- Carborundum Sintered α-SiC (1977 vintage)
- Kyocera SC-201, Sintered α-SiC (1980 vintage)
- Carborundum 1981 SASC (Hexoloy SX-05)
- ESK Sintered α-SiC

G. Silicon-Densified SiC (Siliconized)
- Norton NC-435 Si/SiC
- UKAEA/BNF Refel Si/SiC (diamond-ground and as-processed)
- Norton NC-430 Si/SiC
- Coors Si/SiC (1979, SC-1)
- Coors Si/SiC (1981 and 1982, SC-2)
- General Electric Silcomp Si/SiC (Grade CC)

TABLE 1 (concluded)

H. SiC Coatings
    - Chemetal/SFL CNTD-SiC Coating/Graphite Substrate

I. SiAlON Materials
    - AFML/General Electric GE-129 SiAlON
    - AFML/General Electric GE-130 SiAlON
    - BuMines $ZrSiO_4$-doped Sintered SiAlON (3 compositions)
    - BuMines Y_2O_3-doped Sintered SiAlON

J. Oxide Materials
    - Corning Glass Pyroceram 9606 Glass-Ceramic
    - AFML Zyttrite ZrO_2 (YSZ)
    - Zirconia Ceramics from Australian, German, Japanese, and Domestic Sources
    - Ceres/NRL Single-Crystal Zirconia

TABLE 2. ADDITIONAL MATERIALS INFORMATION

Material	Sample ID Code, 1st letter	Bulk Density, g cm^{-3}	Surface Finish, µin. rms	Comments[a]	Date Received
1 Norton NC-132 HP-Si_3N_4 (1% MgO)	1	3.18	12	Diamond ground at Norton	11/75
2 Norton NCX-34 HP-Si_3N_4 (8% Y_2O_3)	5	3.37	15	Diamond ground at Bomas	8/77
3 Harbison-Walker HP-Si_3N_4 (10% CeO_2)	H	3.34	5	Supplied diamond ground	7/77
4 Kyocera SN-3 HP-Si_3N_4	A	3.06	8	Supplied diamond ground	2/77
5 Norton NC-435 Si/SiC	4	2.96	5	Diamond ground at Norton	1/76
6 Carborundum Sintered α-SiC	8	3.16	10	Supplied diamond ground	3/77
7 General Electric Sintered SiC	G	3.04	12	Supplied diamond ground	2/77
8 UKAEA/BNF Refel Si/SiC	U	3.11	8	Supplied diamond ground (designated batch 1 in material ID notation)	5/77
9 Kyocera SN-201 Sintered Si_3N_4	B	3.00	15	Supplied diamond ground	2/77
10 Kyocera SN-205 Sintered Si_3N_4	2	2.81	20	Supplied diamond ground	2/77
11 Norton NC-350 RS-Si_3N_4	3	2.41-2.55	6	Diamond ground at Norton	1/76

Material	Sample ID Code, 1st letter	Bulk Density, g cm^{-3}	Surface Finish, μin. rms	Comments[a]	Date Received
12 Kawecki-Berylco RS-Si$_3$N$_4$	K	2.35-2.53	20-60	Supplied as-fabricated, diamond ground as reqd. to meet dimension spec.	6/76
13 AiResearch (slip-cast) RS-Si$_3$N$_4$ (RBN-101)	7	2.85	16	Diamond ground at Bomas	8/77
14 Ford (injection molded) RS-Si$_3$N$_4$	F	2.75	29	Supplied as-sintered	8/77
15 Raytheon (isopressed) RS-Si$_3$N$_4$	6	2.43	10-30	Supplied rough ground; ultrasonic machined at Bullen; finish diamond ground at IITRI.	3/77
16 Indussa/Nippon Denko RS-Si$_3$N$_4$	N	2.08	110	Diamond ground at AFML (very porous material)	7/77
17 Naval Res. Lab. HP-Si$_3$N$_4$ + 4-16 w/o ZrO$_2$	R	3.1-3.4	8	Supplied diamond ground	3/77
18 Ceradyne Ceralloy 147A HP-Si$_3$N$_4$ (1% MgO)	X	3.22	20	Diamond ground at Bomas	11/77
19 Ceradyne Ceralloy 147Y HP-Si$_3$N$_4$ (15% Y$_2$O$_3$)	Z	3.37	18	Diamond ground at Bomas	11/77
20 Ceradyne Ceralloy 147Y-1 HP-Si$_3$N$_4$ (8% Y$_2$O$_3$)	Y	3.29	15	Diamond ground at Bomas	11/77
21 GTE Sylvania Sintered Si$_3$N$_4$ (6% Y$_2$O$_3$)	C	3.23	12	Supplied diamond ground	6/78
22 AiResearch (injection molded) RS-Si$_3$N$_4$ (RBN-122)	9	2.66	50-80	Supplied as processed	11/77

	Material	Sample ID Code, 1st letter	Bulk Density, g cm^{-3}	Surface Finish, μin. rms	Comments[a]	Date Received
23	Norton NC-350 RS-Si$_3$N$_4$	3	2.38	22	As-nitrided (designated batch 6 in IITRI sample ID notation)	6/77
24	UKAEA/BNF Refel Si/SiC (as-processed)	U	3.09	40-70	Supplied as-processed (designated batch 2)	11/77
25	Ceradyne Ceralloy 146A HP-SiC (2% Al$_2$O$_3$)	V	3.22	18	Diamond ground at Bomas	11/77
26	Ceradyne Ceralloy 146I HP-SiC (2% B$_4$C)	W	3.21	15	Diamond ground at Bomas	11/77
27	Norton NC-203 HP-SiC (2% Al$_2$O$_3$)	E	3.32	10	Diamond ground at Bomas	9/78
28	Annawerk Ceranox NR-115 H RS-Si$_3$N$_4$	I	2.44-2.68	50-100	As-nitrided	10/78
29	Fiber Materials, Inc. HP-Si$_3$N$_4$ (4% MgO)	J	3.16-3.2	15	Diamond ground at Bomas	10/78
30	Associated Engineering Developments Nitrasil RS-Si$_3$N$_4$	P	2.5-2.6 2.6	25, 100 15	Supplied diamond ground and as-nitrided, 4 batches; batch 5, diamond-ground at Bomas	10/78 10/80
31	Norton NC-350 RS-Si$_3$N$_4$	3	2.4	9	Supplied diamond ground (designated batch 9)	2/79
32	Harbison-Walker HP-Si$_3$N$_4$ (10% CeO$_2$)	H	3.38	16	Supplied diamond ground (designated batch 2)	2/79
33	Norton NC-430 Siliconized SiC	O	3.1	5	Diamond ground at Bomas	3/79

	Material	Sample ID Code, 1st letter	Bulk Density, g cm^{-3}	Surface Finish, μin. rms	Comments[a]	Date Received
34	AiResearch Sintered Si$_3$N$_4$ (8% Y$_2$O$_3$, 4% Al$_2$O$_3$)	L	3.1	30-50	Injection molded. Tested as-nitrided	7/79
35	Rocketdyne SN-50 Sintered Si$_3$N$_4$ (Y$_2$O$_3$, Al$_2$O$_3$)	RC	3.25	35	Injection molded. Tested as-nitrided	11/79
36	Rocketdyne SN-46, SN-104 Sintered Si$_3$N$_4$ (Y$_2$O$_3$, SiO$_2$)	RC	3.4	40	Injection molded. Tested as-nitrided	11/79, 5/80
37	Coors Si/SiC (SC-1)	CG	3.0	10	Supplied diamond-ground	10/79
38	Chemetal CNTD SiC Coating on Graphite Substrate	CS1 CS2 CS3, CS4 CS5, 6, 7	-- -- -- --	40-100 30-50 -- --	UT-22 graphite (Batch 1) SIC-6 graphite (Batch 2) SIC-6 graphite (Batch 3, 4) SIC-6 graphite (Batch 5, 6, 7)	9/79 2/80 9/80 1/81
39	Toshiba HP-Si$_3$N$_4$	TO	3.24	10	Supplied diamond-ground	2/80
40	Toshiba HP-Si$_3$N$_4$	TR	3.20	15	Samples diamond-ground from hot-pressed cylinder liner. Most samples contained bonded joint region	4/80
41	Georgia Tech RS-Si$_3$N$_4$	GT	2.5-2.6	10	Diamond-ground at Bomas	4/80
42	CSIRO Transformation-Toughened ZrO$_2$	ZT	5.7	20	Supplied diamond-ground	7/80
43	Westinghouse HP-Si$_3$N$_4$	WB	3.2-3.3	10	Diamond-ground at Bomas	9/80
44	Kyocera SC-201 Sintered SiC	KK	3.14	30	Supplied diamond-ground	9/80

Materials Evaluation 17

	Material	Sample ID Code, 1st letter	Bulk Density, g cm^{-3}	Surface Finish, μin. rms	Comments[a]	Date Received
45	AME RS-Si$_3$N$_4$	BI1	2.1	130	Supplied diamond-ground	10/80
46	AiResearch RBN-104 RS-Si$_3$N$_4$	BI2	2.8	60	Supplied diamond-ground	10/80
47	Battelle HIP-Si$_3$N$_4$	BW	3.24	14	Supplied diamond-ground	1/80
48	NASA/AVCO/Norton HP-Si$_3$N$_4$ (10% ZrO$_2$)	NA	3.36	17	Diamond-ground at Bomas	6/81
49	Carborundum SASC (Hexoloy SX-05)	8H2	3.1	70	Injection molded, as-fired	8/81
50	Coors Si/SiC (SC-2)	CG2	3.1	30	Supplied diamond-ground	10/81
51	Coors Si/SiC (SC-2)	CG3	3.1	-	Supplied diamond-ground	12/82
52	General Electric Silcomp Si/SiC (Grade CC)	GS1	2.9	~12	Diamond-ground at Bomas	4/82
53	ESK Sintered α-SiC	ES1	3.1	12	Supplied diamond-ground	9/82
54	BuMines Sintered SiAlON (Y$_2$O$_3$)	UH1	3.04	~10	Diamond-ground at Bomas	10/82
55	BuMines Sintered SiAlON (ZrSiO$_4$)	UH2	2.64	~20	Diamond-ground at Bomas	10/82
56	BuMines Sintered SiAlON (ZrSiO$_4$)	UH3	2.63	~60	Diamond-ground at Bomas	5/83
57	BuMines Sintered SiAlON (ZrSiO$_4$)	UH4	2.90	~20	Diamond-ground at Bomas	2/83

[a] All diamond-machined specimens were longitudinally ground (i.e., parallel to the tensile axis of the test bar). Norton refers to Norton Company, Worcester, MA. Bomas refers to Bomas Machine Specialties, Inc., Boston, MA. Bullen refers to Bullen Ultrasonics Company, Eaton, OH.

3. Test Plan

Tables 3 through 5 present the various pre-test and post-test characterization parameters, as well as the general test plan for this program. For any given material, the specific tests that are conducted are a function of number of test samples available, expected properties (determines specific test temperatures), and intended use of the material. In general, tests are conducted at room temperature to establish a baseline, and in the appropriate elevated temperature range where the properties (e.g., strength) are expected to begin to change rapidly (i.e., T ⩾1000°C). All testing is being performed in air. The maximum test temperature is 1500°C. For strength behavior, it is important to recognize this as a screening effort where the typical test sample population is five to ten for any given test temperature. Therefore, only rough estimates of Weibull moduli can be made.

TABLE 3. PRE-TEST CHARACTERIZATION PARAMETERS

- Density-Porosity Characterization
- Microstructural Analysis
- Impurity Content
    - Cations - spectrographic (AFWAL)
    - Oxygen - neutron activation analysis (AMMRC)
- Phase Identification - XRD (AFWAL)
- X-radiographic Inspection (AFWAL)

TABLE 4. POST-TEST ANALYSIS

- Correlation of Properties
- Comparison with Literature Data
- Establish Structure-Property Relations
- Interpretation of Mechanisms
- Failure Mode Analysis, Identification of Critical Flaws by Fractographic Techniques. Performed on virgin and exposed flexural test bars broken at 25°C and at elevated temperatures.

TABLE 5. SCREENING TESTS CONDUCTED[a]

- Flexural Strength-Modulus (4 point), Stress-Strain (25°C, 1000°-1500°C)
- Fracture Toughness
  - Controlled Flaw (25°C, 1000°-1500°C)
- Creep (stepwise) (1300°-1500°C); rate, stress-dependence
- Long Term Oxidation Exposure (100 and 1000 hr at ~1400°C) Residual Strength at 25°C, fracture origins, weight changes, surface scale morphology (optical and SEM), scale products (XRD, XRF)
- Thermal Expansion (25°-1500°C)
- Thermal Diffusivity (25°C, 800°-1500°C)
- Thermal Shock
  - Water Quench/Internal Friction/Residual Strength
  - Analytical Thermal Stress Resistance Parameters
- Dynamic Young's Modulus (room temperature)
- Slow Crack Growth Evaluated on Selected Materials Using Differential Strain Rate Tests
- Stress Rupture

---

[a] Tests conducted in air atmosphere.

4. Materials Characterization

Materials characterization parameters include X-ray phase identification and spectrographic metallic impurity analysis (both performed by AFWAL). Table 6 lists the phases present in the materials studied to date, as determined by X-ray diffraction. Spectrographic impurity analysis results are presented in Table 7. Elements present in quantity <0.2% were detected by emission spectrographic analysis. For elements present in quantities >0.2%, an emission spectrographic technique was used wherein unknown samples were fused, dissolved into solution, and appropriate standard references established. The dilute unknowns were then submitted for standard spectrographic analysis to determine the cation impurity levels. Oxygen analysis of supplied test materials was conducted at USAMMRC, Watertown, Mass. Results obtained to date are presented in Table 8.

TABLE 6. X-RAY DIFFRACTION ANALYSIS RESULTS

Material	Phases Present	
	Major	Minor
Hot-Pressed Si_3N_4		
Norton NC-132 (1% MgO)	β-Si_3N_4	Si_2N_4O
Norton NCX-34 (8% Y_2O)	β-Si_3N_4	$Y_2O_3 \cdot Si_3N_4$
Harbison-Walker (10% CeO_2)	β-Si_3N_4	α-Si_3N_4
Kyocera SN-3 (MgO, Al_2O_3)	β-Si_3N_4	α-Si_3N_4, Si_2N_2O
Ceradyne 147A (1% MgO)	β-Si_3N_4	
Ceradyne 147Y-1 (8% Y_2O_3)	β-Si_3N_4	
Ceradyne 147Y (15% Y_2O_3)	β-Si_3N_4	$YSiO_2N$
Fiber Materials (4% MgO)	β-Si_3N_4	
Toshiba	β-Si_3N_4	
Toshiba (cylinder liner)	β-Si_3N_4	
Westinghouse	β-Si_3N_4	$Y_2Si_2O_7$, α-$Y_2Si_2O_7$, trace $Y_4Si_3O_{12}$
NASA/AVCO/Norton (10% ZrO_2)	β-Si_3N_4	
Battelle HIP-Si_3N_4 (5% Y_2O_3)	β-Si_3N_4	$Y_5(SiO_4)_3N$
Reaction Sintered Si_3N_4		
Norton NC-350 (1976)	α-Si_3N_4	β-Si_3N_4
KBI	a	a
KB-I	α-Si_3N_4	β-Si_3N_4
Ford (IM)	α-Si_3N_4	β-Si_3N_4
AiResearch (SC)	a	a
Raytheon (IP)	α-Si_3N_4	β-Si_3N_4
Indussa/Nippon Denko	α-Si_3N_4	β-Si_3N_4
AiResearch (IM)	α-Si_3N_4	β-Si_3N_4
Norton NC-350 (1977)	α-Si_3N_4	β-Si_3N_4
Norton NC-350 (1979)	α-Si_3N_4	β-Si_3N_4
Annawerk Ceranox	α-Si_3N_4	β-Si_3N_4
AED Nitrasil		
Batch 1	α-Si_3N_4	β-Si_3N_4
Batch 2	α-Si_3N_4	β-Si_3N_4
Batch 3	α-Si_3N_4	β-Si_3N_4
Batch 4	a	a
Batch 5	α-Si_3N_4	β-Si_3N_4
Georgia Tech	a	a
AME	α-Si_3N_4	β-Si_3N_4
AiResearch RBN-104	α-Si_3N_4	β-Si_3N_4

TABLE 6 (cont.)

Material	Phases Present	
	Major	Minor
Sintered Si_3N_4		
Kyocera SN-205 (MgO, Al_2O_3)	β-Si_3N_4	α-Si_3N_4
Kyocera SN-201 (MgO, Al_2O_3)	β-Si_3N_4	
GTE (6% Y_2O_3)	β-Si_3N_4	
Rocketdyne SN-50 (Y_2O_3, Al_2O_3)	β-Si_3N_4	
Rocketdyne SN-104 (Y_2O_3, SiO_2)	β-Si_3N_4	
SiC Materials		
Carborundum SSC (1977)	α-SiC (4H, 15R, and/or 21R)	
General Electric SSC (B1)	β-SiC (3C)	
General Electric SSC (B2)	β-SiC (3C)	Trace α-SiC (15R?)
Norton NC-435 Si/SiC	α-SiC	Si
UKAEA BNF Refel Si/SiC	α-SiC (4H, 15R, and/or 21R)	Si
Norton NC-430 Si/SiC	α-SiC (6H, 15R)	Si
Coors Si/SiC (1979)	α-SiC (2H, 4H, 6H, 15R)	β-SiC (3C), Si
Ceradyne 146A HPSC (2% Al_2O_3)	α-SiC (4H, 15R, and/or 21R)	
Ceradyne 146I HPSC (2% B_4C)	α-SiC (4H, 15R, and/or 21R)	
Norton NC-203 HPSC (2% Al_2O_3)	α-SiC (6H, 33R, 51R)	β-SiC (3C)
Chemetal CNTD SiC Coating		
Batch 1	β-SiC (3C)	Trace α-SiC
Batch 2	β-SiC (3C)	Si; trace α-SiC
Kyocera SC-201 SSC	α-SiC (6H, 15R, 21R?)	
Coors Si/SiC (1981)	α-SiC (6H, 4H, 15R, 2H)	β-SiC (3C)
Carborundum SSC (1981)	α-SiC (6H, 4H, 15R)	
G. E. Silcomp Si/SiC	β-SiC (3C)	Si, trace α-SiC
Coors Si/SiC (1982)	α-SiC (6H, 4H, 2H)	Si, trace α-SiC, 21R
ESK SSC	α-SiC (6H, 4H, 2H)	Trace α-SiC (21R)

[a] Approximately equal amounts of α-Si_3N_4 and β-Si_3N_4.

TABLE 7. METALLIC IMPURITY ANALYSIS[a]

Material	Weight Percent Element[b]									Total Wt% Metallic Impurities incl. Trace Elements
	Al	Fe	Mn	Cr	Ca	Mg	B	W	Others	
Hot-Pressed Si₃N₄										
Norton NCX-34 (8% Y₂O₃)	0.3	0.5	0.05	0.02	0.05			2.25	0.15Co	3.6% + 5.8% Y
Harbison-Walker (10% CeO₂)	0.53	0.6			0.1			0.1		1.3% + 7% Ce
Norton NC-132 (1% MgO)	0.17	0.55	0.05	0.02	0.04	0.84		2.1	0.20Co	4%
Kyocera SN-3	2.6	1.4-1.9	0.04	0.1	0.2-0.33	1.9-2.3			0.1Na, 0.23K	7%
Ceradyne (1% MgO)	0.5	0.77	0.014		0.22	0.7				2.5 - 4.6%
Ceradyne (8% Y₂O₃)	1.0	1.4	0.03		0.17	0.1				2.6% + 6% Y
Ceradyne (15% Y₂O₃)	1.0	1.1	0.024		0.18					2.3% + 12.4% Y
Harbison-Walker (10% CeO₂) Batch 2	0.37	0.81			0.11					1.4% + 5.1% Ce (6.2% CeO₂)
Fiber Materials (4% MgO)	0.18-0.23	0.25			0.12-0.18	2.1-2.9				2.7 - 3.6%
Toshiba	1.6	0.04								1.7% + 2.95% Y
Toshiba (cylinder liner)	2.3	0.03								2.4% + 2.35% Y
Westinghouse	0.05	0.71	0.06		0.02			3.6	0.04Ti, 0.25Co	4.78% + 3.1% Y
NASA/AVCO/Norton (10% ZrO₂)	0.10	0.21	0.05		0.1			1.6	0.09Co	2.2% + 1.7% Y + 5.2% Zr
Battelle HIP (5% Y₂O₃)	0.06	0.07								0.2% + 3.15% Y (4% Y₂O₃)
Hot-Pressed SiC										
Ceradyne (2% Al₂O₃)	1.0	0.44			<0.1		0.1	0.43	0.06Co	2.1 - 3.9%
Ceradyne (2% B₄C)	1.5	0.1					0.92			2.5 - 3.3%
Norton NC-203 (2% Al₂O₃) Batch 1	1.5	0.03		0.19				3.8		5.6%
Norton NC-203 (2% Al₂O₃) Batch 2	1.7	0.55		0.13				3.8		6.3%

TABLE 7 (cont.)

Material	Al	Fe	Mn	Cr	Ca	Mg	B	W	Others	Total Wt% Metallic Impurities incl. Trace Elements
SiC Coatings										
Chemetal CNTD SiC, Batch 1										0.05%
Chemetal CNTD SiC, Batch 2										0.09%
Si/SiC										
UKAEA/BNF Refel (diamond ground)		0.27								0.3 + Si
UKAEA/BNF Refel (as-processed)	0.1	0.58								0.6 – 0.7% + Si
Norton NC-435	0.14	0.4	0.05		0.02	0.02				0.65% + Si
Norton NC-430	0.4	0.1								0.55% + Si
Coors Si/SiC (1979)	0.11	0.12								0.35% + Si
Coors Si/SiC (1981)	0.09	0.11						<0.02		0.33% + Si
Coors Si/SiC (1982 SC-2)	0.16	0.12					0.14		0.05Zr	0.57% + Si
GE Silcomp Si/SiC (CC)	0.36	0.17					0.15		0.06Ti	0.80% + Si
Sintered SiC										
Carborundum (α-SiC) (1977)	0.06–0.09	0.18–0.27					0.41	0.07		0.6%
General Electric (β-SiC) (Batch 1)	0.1	0.19–0.29					0.42–0.47	1.1–1.2	0.075Co	1.9 – 2.5%
General Electric (β-SiC) (Batch 2)	0.04	0.27					0.36	2.0		2.4%
Kyocera SC-201	0.11						0.5	<0.02	0.04Ti	0.73%
Carborundum (1981)	0.09	0.03					0.36			0.25%
ESK (α-SiC)	0.57	0.03					0.08		0.07Ti	0.75%

TABLE 7 (cont.)

Material	Al	Fe	Mn	Cr	Ca	Mg	B	W	Others	Total Wt% Metallic Impurities incl. Trace Elements
Sintered Si$_3$N$_4$										
Kyocera SN-201	3.5-3.8	1.3-1.8	0.03	0.1	0.25-0.39	2.1-2.6				6-8.6%
Kyocera SN-205	4.9-5.0	1.4-2.3	0.04	0.1	0.25-0.45	2.4-3.2				5-11%
GTE Sylvania	0.04	0.12							.1Mo	0.4% + 5.6% Y (7.1% Y$_2$O$_3$)
AiResearch (8% Y$_2$O$_3$, 4% Al$_2$O$_3$)	1.2	0.62							.15Mo	2.1% + 6.4% Y (8.1% Y$_2$O$_3$)
Rocketdyne SN-50 (Y$_2$O$_3$, Al$_2$O$_3$)	2.0	0.05						.02	.15Mo	2.3% + 4.5% Y
Rocketdyne SN-104 (Y$_2$O$_3$, SiO$_2$)	0.16	0.04						1.5	0.08Ti 0.10Co	1.9% +10.6% Y
Reaction Sintered Si$_3$N$_4$										
Norton NC-350 (original batch 1)	0.14	0.4	0.05	0.02	0.04	0.01		<0.02		0.7%
KBI	0.15-0.2	0.41-0.61	0.02	<0.1	0.13-0.17	0.02		0.02		0.8-2%
AiResearch (SC)	0.25	1.1								1.4%
Ford (IM)	0.1	1.0	0.1							1.2%
Raytheon (IP)	0.95	1.3								2.3%
Indussa/Nippon Denko	0.43	0.4		0.1	0.25					1.2%
AiResearch (IM)	0.54	1.1	0.016		0.029					1.4-1.7%
Annawerk Ceranox NR-115 H	0.3	0.52	0.06	0.1	0.04					1.1%

TABLE 7 (cont.)

Material	Weight Percent Element[b]									Total Wt% Metallic Impurities incl. Trace Elements
	Al	Fe	Mn	Cr	Ca	Mg	B	W	Others	

Reaction Sintered Si_3N_4 (cont.)

Material	Al	Fe	Mn	Cr	Ca	Mg	B	W	Others	Total
Norton NC-350 (Batch 6)	0.16	0.32								0.6%
Norton NC-350 (Batch 9)	0.16	0.27								0.5%
AED Nitrasil (Batch 1)	0.75	0.64			0.26					1.8%
AED Nitrasil (Batch 2)	0.72	0.58			0.27					1.75%
AED Nitrasil (Batch 3)	0.22	0.34								0.7%
AED Nitrasil (Batch 4) Georgia Tech	0.24 0.24	0.44 0.28								0.8% 0.7%
AED Nitrasil (Batch 5)	0.49	0.37	0.02		0.21			<0.02	0.04Ti	1.2%
AME	0.72	0.44			0.18			<0.02	0.04Na, 0.06Ti, 0.02V	1.6%
AiResearch RBN-104	0.17	1.3						<0.02	0.02Ni, 0.04Ti, 0.04Mo, 0.04V	1.7%

[a] Emission spectrographic analysis for elements present in quantity <0.2%; quantitative wet chemical analysis for elements >0.2%. Does not include oxygen and unreacted silicon impurities.
[b] Elements present in amount <0.1 wt% usually not recorded, but appear as trace elements in the summed total of metallic impurities

TABLE 8. OXYGEN CONTENT OF CERAMIC TEST MATERIALS

Material	Oxygen, wt%[a]
A. Hot Pressed Si_3N_4	
• Norton NC-132 HP Si_3N_4 (1% MgO)	3.33
• Norton NCX-34 HP-Si_3N_4 (8% Y_2O_3)	3.33
• Harbison-Walker HP-Si_3N_4 (10% CeO_2)	2.81-2.95
• Kyocera SN-3 HP-Si_3N_4 (4% MgO, 5% Al_2O_3)	9.97
• Ceradyne Ceralloy 147A, HP-Si_3N_4 (1% MgO)	1.72
• Ceradyne Ceralloy 147Y, HP-Si_3N_4 (15% Y_2O_3)	5.16
• Ceradyne Ceralloy 147Y-1, HP-Si_3N_4 (8% Y_2O_3)	4.43
• Fiber Materials, Inc. HP-Si_3N_4 (4% MgO)	2.51
• Toshiba (4% Y_2O_3, 3% Al_2O_3)	2.26
• Toshiba (3% Y_2O_3, 4% Al_2O_3, SiO_2)	4.29
• Westinghouse (4% Y_2O_3, SiO_2)	3.16
B. Sintered Si_3N_4	
• Kyocera SN-205 Sintered Si_3N_4 (5% MgO, 9% Al_2O_3)	8.94
• Kyocera SN-201 Sintered Si_3N_4 (4% MgO, 7% Al_2O_3)	6.20
• GTE Sylvania Sintered Si_3N_4 (6% Y_2O_3)	3.04
• AiResearch (IM) Sintered Si_3N_4 (8% Y_2O_3, 4% Al_2O_3)	5.57
• Rocketdyne SN-50 (6% Y_2O_3, 4% Al_2O_3)	4.81
• Rocketdyne (14% Y_2O_3, 7% SiO_2)	7.42
C. Reaction Sintered Si_3N_4	
• Norton NC-350 RS-Si_3N_4 (1976)	0.84
• Kawecki-Berylco RS-Si_3N_4	1.14
• Ford Injection Molded RS-Si_3N_4	1.39
• AiResearch Slip-Cast RS-Si_3N_4 (Airceram RBN-101)	1.26
• Raytheon Isopressed RS-Si_3N_4	1.65

TABLE 8 (cont.)

Material	Oxygen, wt%[a]
C. Reaction Sintered Si_3N_4 (cont.)	
• Indussa/Nippon Denko RS-Si_3N_4	0.85
• AiResearch Injection Molded RS-Si_3N_4 (Airceram RBN-122)	1.75
• Norton NC-350 RS-Si_3N_4 (1977 As-fired)	1.02
• Norton NC-350 RS-Si_3N_4 (1979 Vintage)	0.89
• Annawerk Ceranox NR-115H RS-Si_3N_4	2.29
• Associated Engineering Development, Ltd. Nitrasil RS-Si_3N_4	1.15
• Georgia Tech RS-Si_3N_4	1.74
• AME RS-Si_3N_4	2.66
• AiResearch RBN-104 RS-Si_3N_4	1.48
D. SiC Materials	
• Norton NC-435 Siliconized SiC	0.34
• General Electric Boron-Doped Sintered β-SiC	0.02-0.10
• Carborundum Sintered α-SiC	0.01
• UKAEA/British Nuclear Fuels Refel Si/SiC (siliconized)(diamond ground and as-processed)	0.01-0.02
• Ceradyne Ceralloy 146A, HP-SiC (2% Al_2O_3)	0.49
• Ceradyne Ceralloy 146I, HP-SiC (2% B_4C)	1.28
• Norton NC-203 HP-SiC ($\sim$2% Al_2O_3)	1.27-1.31
• Norton NC-430 Si/SiC	0[b]
• Coors Siliconized SiC	0.03
• Chemetal CNTD SiC Coating	0.05
• Kyocera Sintered α-SiC	0.05
• ESK Sintered α-SiC	0.05
• GE Silcomp Si/SiC Composite	c
• Coors 1982 SC-2 Si/SiC	c

[a] Measured at USAMMRC by neutron activation analysis.
[b] Below limit of detection.
[c] Submitted to AMMRC for measurement.

5. Test Methodology

The test methodology employed on this program follows standard methods and procedures that are used widely throughout the technical ceramic industry.

5.1 REFLECTED LIGHT MICROSCOPY

The goal of microstructural analysis is to reveal the grain size and shape, the amount and distribution of any porosity, and the existence of any highly reflective metal inclusions (impurity particles). This is done by reflected light examination of as-received samples that have been polished and etched to reveal the microstructural features.

Samples of the SiC and Si_3N_4 materials were prepared for optical microscopy by rough grinding through 220, 360, and 600 grit diamond-bonded metal disks using water as a lubricant.[a] They were ultrasonically cleaned between every stage of preparation. Rough polishing was done on chemotextile lap coverings through 9, 6, and 1 μm diamond pastes using a water-base extender. In general, a 0.25 μm diamond micropolishing stage concluded the preparation. However, a final polishing with 0.3 μm α-Al_2O_3 instead of the diamond paste was found to be useful for some of the silicon-densified forms of SiC. The various etching procedures used for the hot-pressed, sintered, and siliconized forms of SiC are summarized in Table 9. The various etching procedures used for the hot-pressed and sintered silicon nitride materials are summarized in Table 10. Etched samples were cleaned and examined with a Leitz MM-5px metallograph.

---

[a]Procedures for the zirconia materials are discussed in Section 12.

TABLE 9. SUMMARY OF MICROSTRUCTURAL ANALYSIS ETCHING PROCEDURES
FOR HOT-PRESSED, SINTERED, AND SILICONIZED SiC

Material	Etching Procedure
Hot-Pressed and Sintered α-SiC	Boiling Murikami's reagent (60 g KOH + 60 g $K_3Fe(CN)_6$ + 120 ml H_2O) 3 to 10 min to preferentially reveal α-α and α-β grain boundaries.[19]
Sintered β-SiC	A fused salt mixture of 90 wt% KOH + 10 wt% KNO_3 at 450°C, 30 sec to 1.5 min to reveal β-β grain boundaries.[19] May be used in conjunction with the Murikami's reagent etch.
Siliconized SiC	An electrolytic etch of 20% KOH in distilled water using direct current, 6 V at 1 amp for 2 to 4 min.[20]

TABLE 10. SUMMARY OF MICROSTRUCTURAL ANALYSIS ETCHING PROCEDURE
FOR HOT-PRESSED AND SINTERED Si_3N_4

a) A chemical etch of 7:2:2 concentrated H_2SO_4:HF:NH_4F solution in a Pt container at 300°C for times from 0.5 to 11 min.[21]

b) HCl swabbed onto surface at room temperature for normally 20 to 60 sec (up to 5 min for some materials).[22]

c) Boling phosphoric/sulfuric acid (3:1) for 6 hr.[23]

d) A solution of 100 ml H_2O + 8 ml HF + 8 g $NaHF_2$ + 1 ml H_2O_2 at 80°C for 1 to 1.5 min.[24]

5.2 FLEXURAL STRENGTH AND ELASTIC MODULUS

Flexure strength was determined in the quarter 4-point configuration on test samples that were of nominal dimensions 1/8 x 1/4 x 2 1/4 inch. Test samples were usually supplied diamond ground or as-nitrided by the manufacturer (refer to Table 2). All samples were corner chamfered. All machined samples were diamond ground in the longitudinal direction, i.e., parallel to the tensile axis of the bend bar.

Unless otherwise indicated, the upper and lower spans were 0.875 and 1.750 in., respectively.* For room-temperature tests, a stainless steel fixture was used. The upper pins were independently free to adjust for any taper across the sample width. The entire upper carrier was free to adjust for any taper along the sample length. Loads were applied in an Instron Universal test machine at a crosshead rate of 0.02 in/min. Elastic modulus was obtained using a 120 Ω foil strain gage applied at the center of the specimen on the tension side. Prior to load application, the alignment was checked visually by inspecting for any light gap between specimen surface and pin. Alignment is also checked by attempting to move the upper pin carriers with gentle pressure of the hand. If the pin is squarely on the specimen surface, movement is not possible.

The outer fiber tensile stress was computed from simple elastic beam theory. Using the nomenclature shown in Figure 1, the lower span is L, the upper span is L/2, and the load applied is P. The outer fiber tensile stress for the bar in bending is

$$\sigma = \frac{Mc}{I} \tag{1}$$

---

*A few samples tested were only 1.25 in. long. The upper and lower spans in these cases were 0.500 and 1.000 inch, respectively.

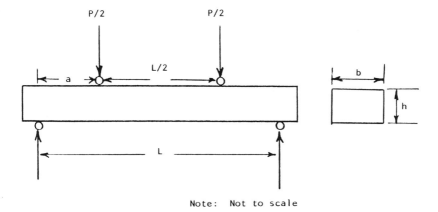

Figure 1. Quarter 4-point flexure configuration.

where σ = outer fiber tensile stress (flexure stress)

M = bending moment = Pa/2

a = L/4 for the quarter 4-point configuration

c = distance from the specimen neutral axis to the outer tensile fiber = h/2

I = moment of inertia of the specimen cross-section about the neutral axis = $bh^3/12$

Substituting:

$$\sigma = \frac{3Pa}{bh^2} \qquad (2)$$

High temperature flexure tests were conducted in an IITRI designed and constructed SiC-element furnace which was rolled in and out of an Instron load frame. The furnace was equipped with a shuttle to move specimens into test position while the furnace was at temperature, and with several access ports to permit manipulation of the specimen and fixtures. The fixturing is unique in that it provides for some degree of self-alignment at temperatures up to 1500°C. The system is shown schematically in Figure 2. Photographs of various parts of the fixturing are shown in Figure 3. Silicon carbide pins are fixed to the saddle fixture (refer to Figure 2). The specimen is pushed onto the saddle fixture using the Al_2O_3 alignment pins to center the specimen. The top load rod/upper load fixture assembly is then lowered into position with the aid of the Al_2O_3 alignment pins. The upper load fixture (upper span) allows alignment in two directions through the use of the key/side plate method of attachment to the top loading rod (see Figure 3). The top loading rod applies a line contact load to the upper fixture load pin. Outside the furnace, alignment is further attained through the use of hemispherical load platens located between the loading rod and the Instron movable crosshead. Fixtures are made with a combination of materials, mostly various commercially available forms of silicon carbide.

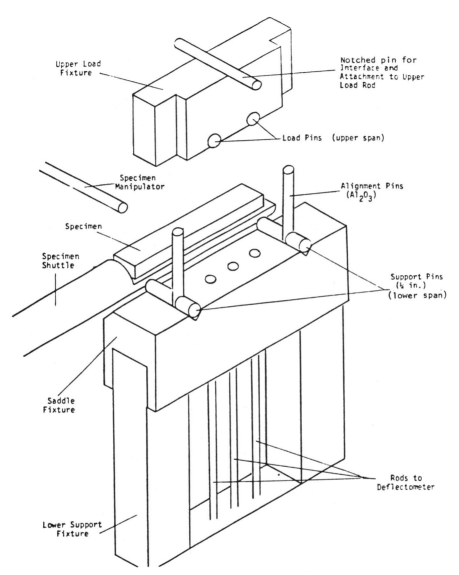

Figure 2. Schematic of high-temperature flexure apparatus.

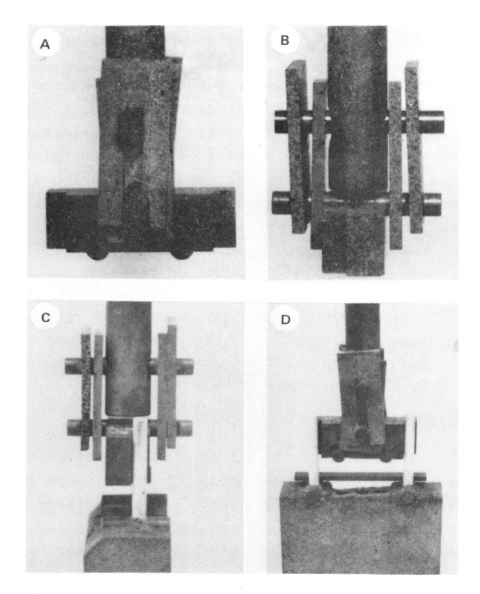

Figure 3. Various views of silicon carbide flexure fixturing.

(a)(b) Front and side views of upper load fixture and loading rod. Load is applied to pin at top of upper load fixture. Keys and side plates allow alignment in two directions while providing attachment to load rod.

(c)(d) Front and side views of top load rod/upper load fixture assembly being lowered onto saddle fixture.

Elevated temperature flexural deformation was recorded continuously using the IITRI-designed and built 3-point electro-mechanical deflectometer shown in Figure 4. The three deflectometer rods (KT-SiC, Al_2O_3, or sapphire) contact the tensile surface of the bend bar within the region of pure bending. Accounting for test fixture deformation is not necessary, since the test sample deformation is measured directly from the relative movement of the center rod with respect to the mean position of the two outer rods, giving $\Delta\delta$, the beam deflection, which is related to the outer fiber tensile strain. From simple beam theory,[25] with reference to Figure 4, the maximum deflection at the center of the beam, δ_{max}, is

$$\delta_{max} = \frac{Pa}{48EI}(3L^2 - 4a^2) \quad (3)$$

The deflection at the outer deflectometer rod, δ_x is

$$\delta_x = \frac{Pa}{12EI}[3Lx - 3x^2 - a^2] \quad (4)$$

The sample deformation $\Delta\delta = \delta_{max} - \delta_x$ is the difference in these two deflections

$$\Delta\delta = \frac{Pa}{4EI}[\tfrac{1}{4}L^2 - Lx + x^2] \quad (5)$$

The heart of this deflectometer is a DC-DC displacement transducer* having 10^{-5} in. resolution. This device is an integrated package consisting of a precision linear variable differential transformer, a solid state oscillator and a phase-sensitive demodulator. The output of this high linearity, high resolution, high sensitivity transducer is a DC voltage proportional to the displacement of the axial core (center deflectometer rod) within a coil assembly (attached to outer two

---

*Trans-Tek Series 240.

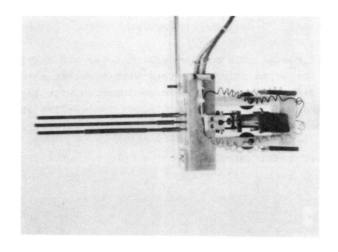

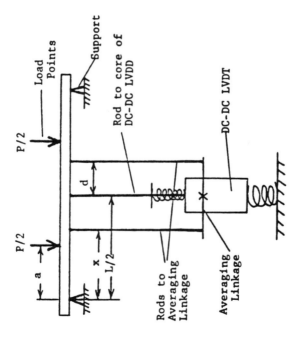

Figure 4. Electromechanical deflectometer for direct measurement of flexural specimen strain.

deflectometer rods). The entire deflectometer was calibrated
with a high-precision micrometer head with resolution to 10^{-5} in.
Use of this deflectometer system in conjunction with the Instron
load cell permits recording of continuous load-deflection curves,
which are then used to construct stress-strain curves.

5.3 CREEP AND STRESS RUPTURE

Creep rates were measured in 4-point bending at temperatures
up to 1500°C in air. The stepwise method was used, with deformation continuously recorded using the 3-point electromechanical
deflectometer described above. Samples were tested at various
incremental stress levels, with enough time spent at any given
level to obtain stage 2 (secondary) creep (maximum of 24 hr at
any stress level). Deadweight loading was employed. Testing
continued for 3-4 stress steps per sample. Testing of approximately three samples in this manner permitted the steady-state
stress dependence of the creep rate to be assessed for each material studied. The interior of the creep test rig, showing the
SiC fixturing, is presented in Figure 5.

Static fatigue tests were conducted on a stress rupture
apparatus with deadweight loading. A photograph is provided in
Figure 6. The sample was loaded at constant stress, and the time
to failure was recorded.

Dynamic fatigue testing has also been used on this program.
These tests are often referred to as differential strain rate
flexure tests, and are conducted on an Instron universal test
machine. The stressing rate (i.e., crosshead speed) was varied
to give a range of sample failure times. The resulting strength-time curve is analyzed like a static fatigue or stress rupture
curve, i.e., by linear least-squares curve fitting analysis.

5.4 DYNAMIC ELASTIC MODULI

A flexural sonic resonance technique was used to measure the
dynamic Young's elastic modulus. The apparatus conforms to that

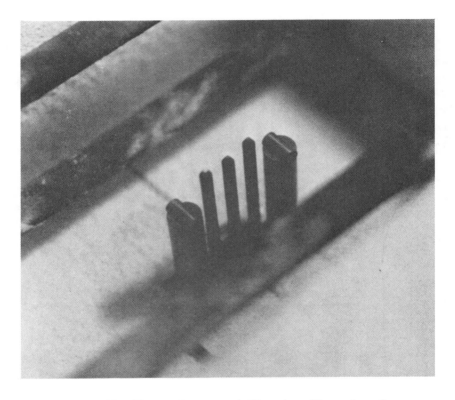

Figure 5. SiC flexural creep deflection fixturing for use at 1500°C in air.

Materials Evaluation   41

Figure 6. Stress-rupture furnaces for use
           at 1500°C in air.

described by Spinner and Tefft,[26] which is the accepted standard of the industry. An oscillator (sonic frequency) signal was fed to a power amplifier, which energized a piezoelectric driver. The driver frequency was measured with a frequency counter. The mechanical vibration was transmitted to and from the sample by suspending it from the driver and pickup transducers. Cotton thread worked well at room temperature while platinum/rhodium wire may be used at elevated temperature. The pickup signal was amplified and detected on a meter and/or as a maximum in the Lissajou pattern on an oscilloscope.

Suspending a flexure bar of nominal dimensions 0.090 x 0.250 x 2.500 in. from points just adjacent to the flexural nodal points permitted detection of the flexural resonant frequency, and subsequent computation of the dynamic Young's moduli from the relation[26]

$$E = \frac{0.94642}{386.09} \frac{\rho L^4 f^2 T}{t^2} \qquad (6)$$

where E = Young's elastic modulus for flexural resonance of a prism of rectangular cross-section, psi

ρ = density, lb in.$^{-3}$

L = sample length, in.

f = flexural resonant frequency (fundamental mode), Hz

t = sample thickness, in.

T = correction and shape factors given by Spinner and Tefft.[26]

Similarity, if a wider sample is used, e.g., .090 x 1.000 x 2.500 in., the sample can be suspended from opposite corners, and torsional as well as flexural vibrations induced.[26] This is mentioned because it is a relatively easy way to obtain Poisson's ratio for ceramics. The shear modulus is given by the relation

$$G = \frac{4f^2 \rho R L^2}{386.09} \qquad (7)$$

where   G = shear modulus for torsional resonance

f = torsional fundamental resonant frequency

R = shape factor[26]

Poisson's ratio, μ, is computed from the Young's and shear moduli using the relation:

$$\mu = \frac{E}{2G} - 1 \qquad (8)$$

5.5 THERMAL SHOCK/INTERNAL FRICTION

Thermal shock resistance was determined on this program by the water quench method, with the initiation of thermal shock damage being detected by internal friction measurement. This technique was chosen not in an attempt to simulate in-service engine conditions, but rather as a relative ranking of candidate materials in severe thermal down-shock. In conducting this test, internal friction was measured before and after water quench from successively higher temperatures using the flexural resonant frequency Zener bandwidth method. A marked change in internal friction (specific damping capacity) indicated the onset of thermal shock damage (i.e., thermal stress-induced crack initiation). This defined the critical quench temperature difference, ΔT_c, which was compared to analytical thermal stress resistance parameters.

Zener[27] and others[28-30] have provided excellent reviews of internal friction. Hookean elastic theory implies a direct and instantaneous linear relation between low level force application and resultant deformation. However, as the rate of loading and unloading is increased, there appears a phase lag between stress and strain which results in the absorption of energy. This time-dependent elastic behavior is termed anelasticity. Internal friction may be defined as the amount of energy absorbed during deformation compared to the maximum amount of energy applied initially. As the number and extent of flaws in a body increases, the amount of energy absorbed through flaw surface friction, plastic zone dislocation motion, etc., also increases. Thus, the internal friction is an integrated effect, the measurement of the total flaw spectrum in a material.

On the present program, test samples were suspended from piezoelectric drive and pickup transducers as described above for dynamic elastic modulus measurement. Using a Hewlett-Packard spectrum analyzer,[*] the frequency range just adjacent (above and below) the fundamental resonant frequency was scanned. The Zener bandwidth[27] method involves computing the internal friction, Q^{-1}, from the measured peak width at half maximum amplitude:

$$Q^{-1} = \frac{\Delta f}{\sqrt{3}\, f} \qquad (9)$$

where Q^{-1} = internal friction

f = resonant frequency

Δf = peak width at half amplitude

The concept is illustrated in Figure 7. This method is useful for measuring the range of internal friction normally found in ceramic materials (10^{-2} to 10^{-6}).

5.6 THERMAL EXPANSION

Thermal expansion was measured from ambient room temperature to 1500°C using a NETZSCH automatic recording single pushrod dilatometer. Samples of nominal dimensions 1/4 x 1/8 x 2 in. were temperature cycled in air at 1°C/min from 25° to ~200°C, and at 5°C/min from 200° to 1500°C. The cooling rate was approximately 5°C/min. Sample length changes were continuously recorded during both heating and cooling cycles with a precision LVDT system exhibiting 1 micron resolution. The apparatus is illustrated in

---

[*]Model No. 3580A.

Materials Evaluation 45

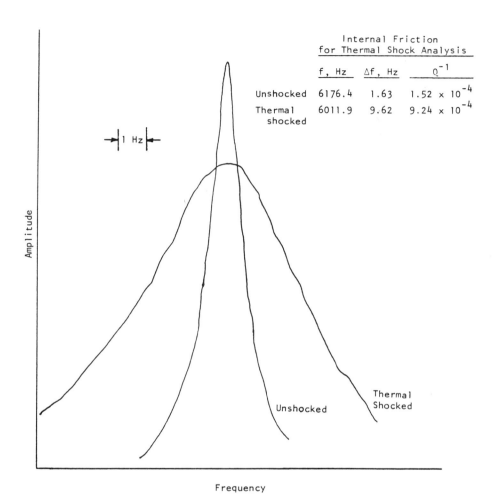

Figure 7. Typical amplitude-frequency curves of unshocked and thermal shocked ceramics for internal friction measurement.

Figure 8. A tungsten standard reference material* was used to calibrate the system. The data were reported as percent linear expansion vs. temperature. From these data, any average or instantaneous coefficient of expansion can be computed.

5.7 THERMAL DIFFUSIVITY

Thermal diffusivity is defined as the ratio of the thermal conductivity to the density-specific heat product. The thermal diffusivity was measured in air at 25°C, and at various temperatures from 1000° to 1500°C. The sample was contained in an IITRI-built molybdenum wire-wound zirconia tube furnace with a porous zirconia sample holder. The laser pulse method[31] was employed. The front face of the disk-shaped sample (9/16 in. dia. x 0.050 in. thick) was irradiated with a single pulse of laser energy (25 joule pulse at 6943 Å, 500 μsec pulse duration),§ and the resulting sample back face temperature transient recorded. A liquid nitrogen-cooled indium antimonide detector** was used to detect the temperature response at 25°C, and a biased silicon photodiode§§ was used at elevated temperatures. The general apparatus is shown in Figures 9 and 10, and a typical rear face temperature response is shown in Figure 11. Initial time, t_o, for the measurement was determined when lasing occurred through the use of an auxiliary photodiode system monitoring the ruby laser performance.

Thermal diffusivity, α, was computed from the expression:

$$\alpha = \frac{0.1388\ L^2}{t_{1/2}} \qquad (10)$$

---

*NBS SRM NO. 737.

§Korad pulsed ruby laser, Model K1.

**Texas Instruments, ISV 3105

§§EGG SGD-100-A.

Materials Evaluation 47

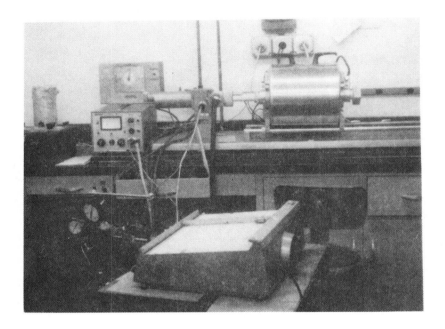

Figure 8. Thermal expansion apparatus.

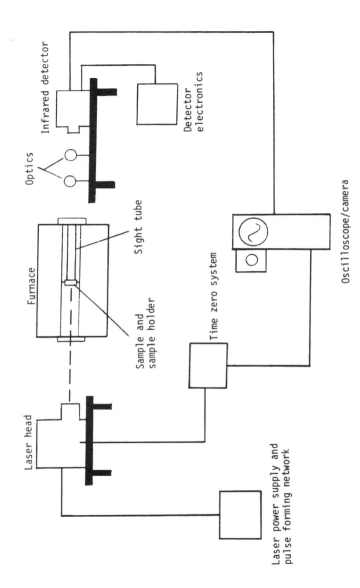

Figure 9. Schematic of laser flash thermal diffusivity method.

Materials Evaluation 49

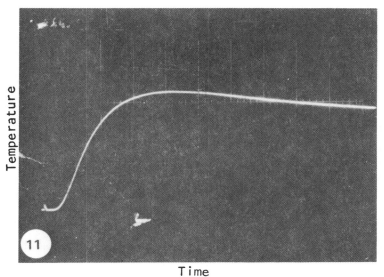

Figure 10. Thermal diffusivity apparatus.

Figure 11. Temperature transient for thermal diffusivity measurements.

where L = sample thickness, cm

$t_{1/2}$ = time for rear face temperature to reach half its maximum value.

If the laser pulse width is not small compared to the characteristic transit time through the sample, a correction must be applied to account for the resulting finite pulse time effect. This usually occurred for high density, high thermal diffusivity silicon carbide tested at low temperature. In that case the finite pulse time correction[32-34] was applied to the data using the relation:

$$\alpha = \frac{(0.3155)L^2}{(2.273)t_{1/2} - \tau} \qquad (11)$$

The term τ is the pulse width, and this relationship applies for $t_{1/2}/t_c < 1.8$, t_c being the sample characteristic rise time:

$$t_c = \frac{L^2}{\alpha \pi^2} \qquad (12)$$

One of the boundary conditions employed to obtain the ideal analytical solution to this experiment is that there are no heat losses from the specimen surfaces that would alter the shape of the temperature transient. This boundary condition is often violated by radiation losses at high temperatures. Taylor[34] discusses a convenient correction for this effect based on Cowan's original work.[35] This correction involves comparison of the shape of the experimentally obtained temperature transient at times out to ~$10t_{1/2}$ to the shape of the ideal rear face temperature transient obtained from the ideal analytical solution (the general shape of the rear face temperature transient is independent of sample thermal diffusivity). The correction was applied to all data generated on this program above 800°C.

6. Microstructure, Room-Temperature Strength, and Elastic Properties

The room temperature strength and fracture mode of silicon-base ceramics are determined by the microstructural features of grain size and shape, the size and distribution of porosity, and the nature of the particular processing defects (or machining flaws) that cause fracture (i.e., the type of defect, and its size and shape). The elastic properties are determined by the nature of the atomic bonding in the basic system.

6.1 Si_3N_4 MATERIALS

Silicon nitride materials are processed by hot-pressing, sintering, and reaction sintering. The highest strengths are achieved with HP-Si_3N_4, the use of pressure-assisted consolidation and oxide additives (MgO, Y_2O_3, CeO_2, ZrO_2) resulting in nearly fully dense materials. Pressureless sintering also involves the use of oxide densification aids. Densities and strengths are somewhat lower than for hot-pressed forms. Reaction sintered Si_3N_4 is processed by nitridation of precompacted silicon powder. This process results in a relatively porous body (up to 20% porosity), with correspondingly lower strength.

6.1.1 Microstructural Features

The microstructures of the various hot-pressed Si_3N_4 materials evaluated on this program are presented in Figure 12. These micrographs were generally obtained from polished/etched sections viewed in reflected light. The micrographs for the various sintered Si_3N_4 materials are presented in Figure 13.[*]

---

[*]The results for the highly developmental sintered SiAlON materials from the U.S. Bureau of Mines are presented in Reference 36.

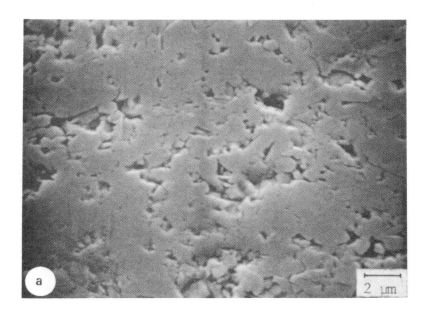

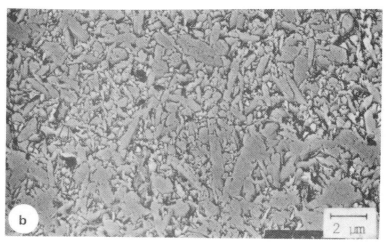

Figure 12. SEM micrographs of polished and etched hot-pressed silicon nitride materials (Etchant: 7:2:2 H_2SO_4/HF/NH_4F etchant shown in Table 10).

(a) Norton NC-132 (1% MgO) etched 6 min.

(b) Replica TEM micrograph of Norton NC-132 (1% MgO). Courtesy of K. S. Mazdiyasni, AFWAL.

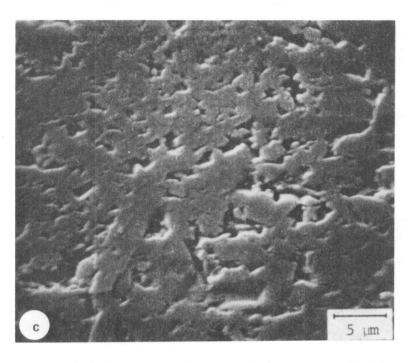

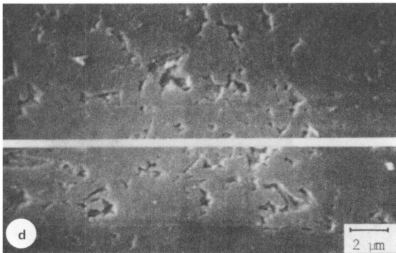

Figure 12 (cont.)
(c) Norton NCX-34 (8% Y_2O_3) etched 6 min.
(d) Harbison-Walker (10% CeO_2) etched 3.5 min.

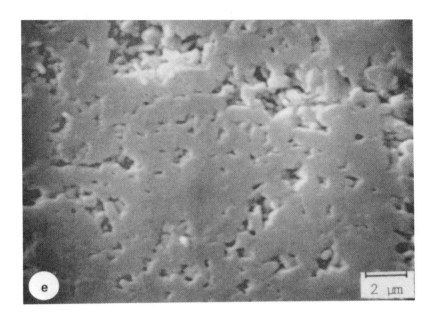

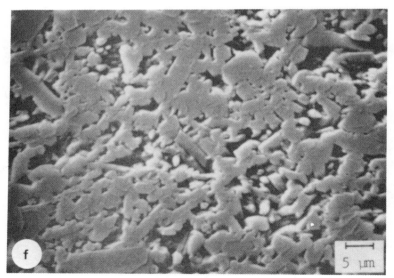

Figure 12 (cont.)

(e) Kyocera SN-3 (4% MgO, 5% Al_2O_3) etched 15 sec at room temperature.

(f) Ceradyne Ceralloy 147A (1% MgO) etched 10 min.

Materials Evaluation 55

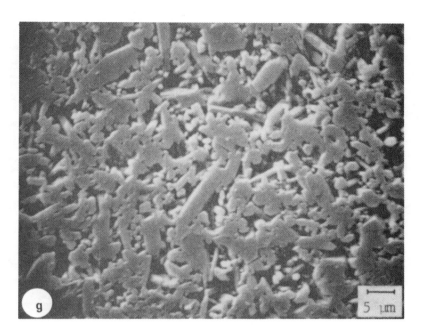

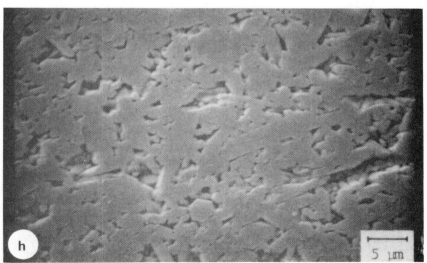

Figure 12 (cont.)

(g) Ceradyne Ceralloy 147Y (15% Y_2O_3)
    etched 6 min
(h) Ceradyne Ceralloy 147Y-1 (8% Y_2O_3)
    etched 7 min

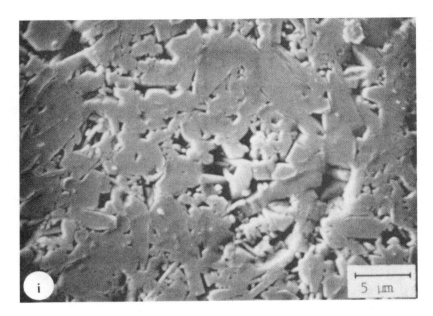

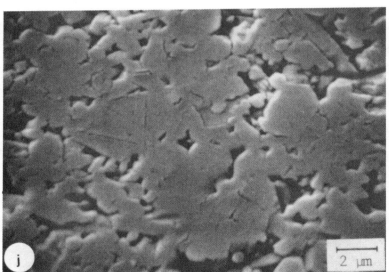

Figure 12 (cont.)

(i) Fiber Materials, Inc., (4% MgO) etched 11 min

(j) Westinghouse (4% Y_2O_3, SiO_2) etched 5 min

Materials Evaluation 57

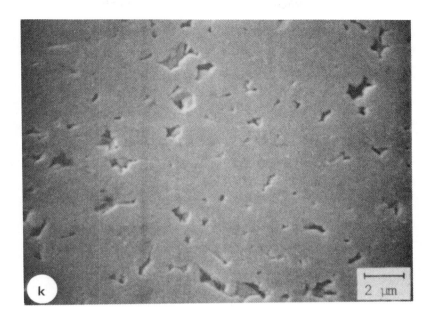

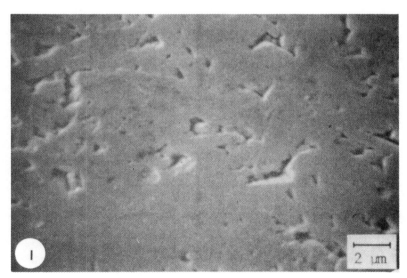

Figure 12 (cont.)

(k) Toshiba (4% Y_2O_3, 3% Al_2O_3)
    etched 0.5 min
(l) Toshiba (3% Y_2O_3, 4% Al_2O_3, SiO_2)
    etched 0.75 min

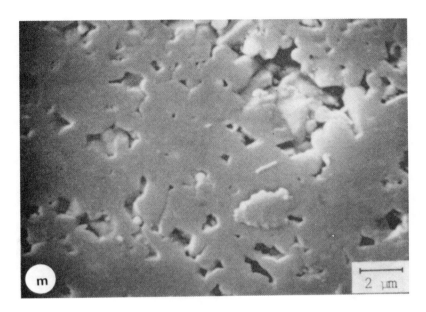

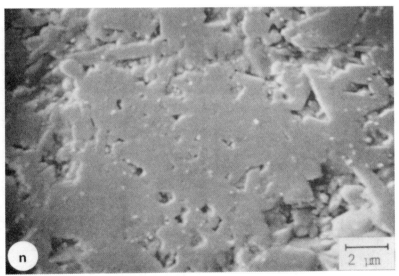

Figure 12 (cont.)

(m) NASA/AVCO/Norton (10% ZrO_2) etched 2.5 min

(n) Battelle HIP-Si_3N_4 (5% Y_2O_3) etched 2.5 min

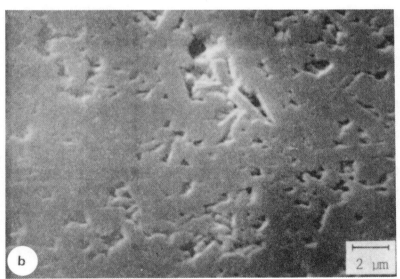

Figure 13. SEM micrographs of etched sintered silicon nitride materials.
(a) Kyocera SN-205 (5% MgO, 9% Al_2O_3) etched 55 sec[a]
(b) Kyocera SN-201 (4% MgO, 7% Al_2O_3) etched 35 sec[a]

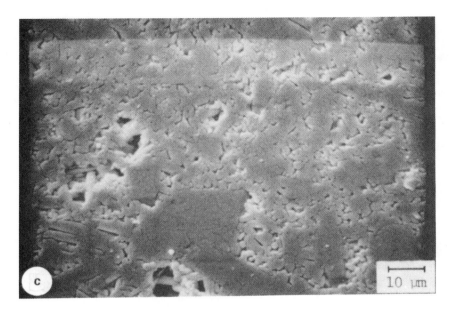

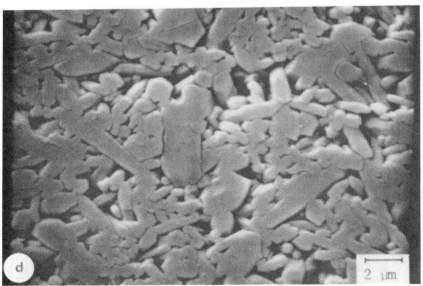

Figure 13 (cont.)
(c) GTE Sylvania (6% Y_2O_3) etched 6 hr[b]
(d) AiResearch (8% Y_2O_3, 4% Al_2O_3) etched 5 min[c]

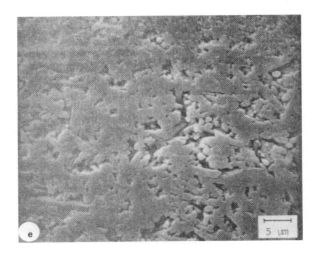

Figure 13 (cont.)

(e) Rocketdyne SN-50 (6% Y_2O_3, 4% Al_2O_3) etched 9.5 min[c]

(f) Rocketdyne SN-104 (14% Y_2O_3, 7% SiO_2) etched 6 min[c]

[a] H_2SO_4/HF/NH_4F shown in Table 10.
[b] HCl etch shown in Table 10.
[c] 3:1 Phosphoric/sulfuric acid etch shown in Table 10.
[d] H_2O/HF/$NaHF_2$/H_2O_2 etch shown in Table 10.

These micrographs reveal grain size and shape, as well as the location of any porosity. The microstructure of reaction-sintered Si_3N_4 materials was evaluated by slightly different means. There is no significant intergranular phase in $RS-Si_3N_4$ that can be readily etched to reveal grain size and shape. Thus, the major features observed in the microstructural analysis of $RS-Si_3N_4$ materials are the size and distribution of porosity and the existence of any free silicon metal. Pores and pore agglomerates are often identified as the fracture origins in $RS-Si_3N_4$. Accordingly, polished sections were prepared of all $RS-Si_3N_4$ materials evaluated on this program. Both reflected light (optical) microscopy and scanning electron microscopy were used to view and record the nature of the microstructure of these materials. The photographic results are compiled in Appendix A.

The results of microstructural analysis for the hot-pressed, sintered, and reaction sintered Si_3N_4 materials (as well as sintered SiAlON materials) are summarized in Table 11. In this tabulation, we include density, mean room-temperature fracture strength, and the average and maximum grain size. The approximate grain size determination was made by an overall visual inspection of the micrographs, and not quantitatively determined by lineal analysis or other statistical techniques. The comments in Table 11 regarding fracture origins were made from the fracture surface analysis of bend bars broken at room temperature.

It is observed that for fully dense $HP-Si_3N_4$, the average grain size typically ranged from 0.5-2 µm. In some materials, occasional grains as large as 17 µm were observed. For sintered Si_3N_4, with usually slightly lower density, the grain size was also typically 0.5-2 µm. The size range of the largest grains observed in these materials was 5-10 µm. These HP- and sintered-Si_3N_4 materials generally exhibit grains that are slightly elongated and randomly oriented. This leads to an interlocking type structure that generally results in higher fracture toughness than in other types of ceramics where the grains are equiaxed.

TABLE 11. MICROSTRUCTURAL ANALYSIS SUMMARY FOR Si_3N_4 MATERIALS

Material	Bulk Density, g cm^{-3}	Mean 25°C Bend Strength, psi	Grain size,[a] μm		Room Temperature Fracture Origins
			Average	Maximum	
Hot-Pressed Si_3N_4					
Norton NC-132 (1% MgO)	3.18	103,100	0.5-1.0	3.0	Primarily machining flaws, other processing defects.
Norton NCX-34 (8% Y_2O_3)	3.37	126,730	0.5-1.8	8.0	Machining flaws and processing defects.
Harbison-Walker (10% Ceria)	3.33	76,780	--	--	Primarily inclusions.
Kyocera SN-3 (4% MgO, 5% Al_2O_3)	3.07	74,860	0.5-2.5	--	Primarily inclusions.
Ceradyne Ceralloy 147A (1% MgO)	3.22	87,090	1.0-5.0	12-17	Primarily inclusions.
Ceradyne Ceralloy 147Y (15% Y_2O_3)	3.37	87,850	0.5-3.0	14	Undetermined.
Ceradyne Ceralloy 147Y-1 (8% Y_2O_3)	3.29	83,250	1.0-3.0	10-12	Machining flaws and processing defects.
Fiber Materials, Inc. (4% MgO)	3.17	66,840	0.5-1.8	5-7	Dark inclusions.
Toshiba (4% Y_2O_3, 3% Al_2O_3)	3.25	105,700	--	--	Dark, shiny inclusions.
Toshiba (3% Y_2O_3, 3% Al_2O_3, SiO_2)	3.20	83,600	--	--	Inclusions.
Westinghouse (4% Y_2O_3, SiO_2)	3.25	90,980	0.5-2.0	--	Dark, shiny inclusions.
NASA/AVCO/Norton (10% ZrO_2)	3.37	91,190	--	--	Dark, shiny inclusions, other processing defects.
Battelle HIP (5% Y_2O_3)	3.25	90,000[c]	--	--	See footnote c
Sintered Si_3N_4					
Kyocera SN-205 (5% MgO, 9% Al_2O_3)	2.81	37,770	0.5-1.0	--	Inclusions and porosity.
Kyocera SN-201 (4% MgO, 7% Al_2O_3)	3.00	49,610	0.5-2.0	4.0	Primarily inclusions, some pores.
GTE Sylvania (6% Y_2O_3)	3.23	78,000	0.5-3.0	5-8	Inclusions, pores and pore/inclusions.
AiResearch (8% Y_2O_3, 4% Al_2O_3)	3.10	79,450	0.5-2.0	6.0	Surface and subsurface porosity.
Rocketdyne SN-50 (6% Y_2O_3, 4% Al_2O_3)	3.25	51,600	0.5-2.0	8-9	Porosity open to tensile surface.
Rocketdyne SN-104, SN-46 (14% Y_2O_3, 7% SiO_2)	3.40	51,600	0.5-1.5	5-6	Porosity open to tensile surface.

TABLE 11 (cont.)

Material	Bulk Density, g cm^{-3}	Mean 25°C Bend Strength, psi	Approximate Pore Size, μm[a,b] Average	Approximate Pore Size, μm[a,b] Maximum	Nature of Porosity	Room Temperature Fracture Origins
Reaction Sintered Si$_3$N$_4$						
1976 Norton NC-350	2.41-2.55	29,450	0.5-3.0	12	Even distribution of pores and pore clusters.	Primarily undetermined, some processing flaws.
Kawecki-Berylco (Batch 2, 1976)	2.35-2.53	19,940	10-15	70	Even distribution of porosity. Fine, evenly distributed free Si.	Particles, pores, some undetermined.
Kawecki-Berylco (Batch 3, 1976)	2.35-2.53	19,940	5-15	30-70	Some large interconnected pores 100 μm. Uneven distribution of free Si.	Particles, pores, some undetermined.
Ford Injection-Molded	2.75	38,180	2-8	100	Avg. clusters 10-20 μm w/large clusters 30-100 μm. Some free Si.	Processing defects, inclusions and pores.
AiResearch Airceram RBN-101 (SC)	2.85	37,920	2-5	35	Angular pores. Avg. clusters 12-25 μm. Some free Si evenly distributed.	Primarily inclusions.
Raytheon (Isopressed)	2.43	21,550	0.5-5.0	25	Even porosity. Avg. cluster 5-10 μm.	Processing defects and undetermined.
Indussa/Nippon Denko	2.08	10,580	20-40	180	Much large, interconnected porosity.	Some pores, some undetermined.
AiResearch Airceram RBN-122 (IM)	2.66	32,580	0.5-3.0	20	Avg. cluster size 5-15 μm. Some free Si unevenly distributed.	Inclusions, some with pores.
1979 Norton NC-350	2.40	37,560	0.5-5.0	20	Large pores 5-10 μm with clusters 15-22 μm. No obvious free Si visible.	Primarily inclusions.
Annawerk Ceranox NR-115H	2.44-2.68	28,870	2-10	60	Many large pores and clusters 30-60 μm. Some free Si visible.	Inclusions.
AED Nitrasil Batch 1 1978	2.52	29,920	2-10	85	Many clusters 20-40 μm unevenly distributed. Little free Si.	Inclusions, some with pores.
2 1978	2.50	27,400	2-10	50	Avg. clusters 25-35 μm with some free Si unevenly distributed.	Primarily inclusions, some pores.
4 1978	2.60	30,730	1-5	50	Avg. clusters 20-40 μm evenly distributed.	Inclusions and inclusions with pores.
5 1980	2.62	26,080	2-10	70	Unevenly distributed clusters 20-50 μm. uneven free Si.	Primarily dark inclusions.
Georgia Tech	2.50-2.60	26,220	0.5-5	18	Even, fine porosity. Fine, even, free Si. Clusters of pores 5-10 μm.	Primarily light inclusions.
AME	2.10	7,470	--	85	Much interconnected porosity. Large areas of free Si unevenly distributed.	Undetermined due to material porosity.
AiResearch RBN-104	2.80	40,270	1-5	24	Fine, even porosity with clusters 8-10 μm.	Subsurface pores and inclusions.
1977 Norton NC-350	2.38	29,450	0.5-3.0	12	Evenly distributed clusters 5-12 μm.	Pores, pore agglomerates.

TABLE 11 (concluded)

Material	Bulk Density, g cm^{-3}	Mean 25°C Bend Strength, psi	Grain size,[a] μm Average	Grain size,[a] μm Maximum	Approximate Pore Size,[a,b] μm Average	Approximate Pore Size,[a,b] μm Maximum	Nature of Porosity	Room-Temperature Fracture Origin
US Bureau of Mines[c] Sintered SIALON[d]								
● Billet 1 (5% ZrSiO$_4$)	2.900 (2.08-3.06)	26,660	<1	--	~5	20	fine interconnected pore structure.	subsurface and surface-connected pores.
● Billet 2 (5% Y$_2$O$_3$)	3.038	35,330	<1	--	<10	30-100	generally discrete pore structure with wide size variation.	large subsurface pores.
● Billet 4 (5% ZrSiO$_4$)	2.632	13,190	<1	--	5-10 agglomerated	50 single pores	much interconnected porosity.	mostly undetermined due to ill-defined fracture features.
● Billet 5 (5% ZrSiO$_4$)	2.646	17,160	<1	--	2-10	20-30, agglomerated	much interconnected porosity.	mostly undetermined; some large pores and large grains identified as fracture initiation sites.

[a] From overall visual inspection of micrographs; not statistically determined by lineal analysis, etc.
[b] Void features with rounded corners were judged to be porosity; void features with sharp/angular corners were judged to be polishing pullouts.
[c] See Reference 5.
[d] Refer to Reference 36 for details of processing and compositional modifications. All SIALON compositions were nominally 80% Si$_3$N$_4$-20% Al$_2$O$_3$, with ~5% additive (either Y$_2$O$_3$ or ZrSiO$_4$) to promote sintering.

Pore size, rather than grain size, was characterized for the reaction-sintered silicon nitride materials since porosity is generally the strength-limiting feature of the microstructure of $RS-Si_3N_4$. With reference to Table 11, it is seen that a wide range of average and maximum porosity was observed. The materials with the finest pore structures were Norton NC-350 and AiResearch RBN-122. Average pores ranged in size from 0.5-3.0 μm, while the largest pores were 12-20 μm in size. In some of the other $RS-Si_3N_4$ materials, average pore sizes were as great as 15 μm, and the maximum pores approached 100 μm in size.

6.1.2 Flexural Strength/Fracture Sources and Elastic Properties

The room temperature 4-point bend strength and elastic moduli (static and dynamic) of all Si_3N_4 materials evaluated to date are tabulated in Tables 12 and 13. Strength data are plotted as a function of bulk density and volume fraction porosity in Figures 14 and 15, respectively. Density (and thus strength) generally increases with respect to processing method in the following progression: reaction sintered with few additives, reaction sintered with iron additives, sintered with oxide additives, hot-pressed with magnesia additive, and finally hot-pressed with yttria additive. A least-squares regression of the linearized form of the exponential strength-porosity relation $\sigma = \sigma_0 e^{-bP}$ gives

$$\sigma = 87.1e^{-5.55P}$$

as shown in Figure 15. The strength behavior of the thirty-nine (39) hot-pressed, sintered, and reaction bonded Si_3N_4 materials is well described by the strength-porosity data fit. This empirical strength-porosity relation is not strictly valid at low porosity, however, even though it accurately describes the strength of the high density hot-pressed materials in this region. This is the case because the fracture strength of the hot-pressed materials is not expected to be related directly to porosity as

TABLE 12. SUMMARY OF ROOM-TEMPERATURE POROSITY, STRENGTH, AND ELASTIC MODULUS DATA FOR VARIOUS Si_3N_4 MATERIALS

Symbol in Figs. 14-16	Material	Bulk Density g cm^{-3}	Theoret. Density[a]	P Volume Fraction Porosity	4-Point Flexure Strength ksi	Static Young's Modulus[b] 10^6 psi
○	Norton NC-132 HP-Si$_3$N$_4$ (MgO)	3.177	3.2	.007	90.91	48.4
○	Norton NC-132 HP-Si$_3$N$_4$ (MgO)	3.186	3.2	.004	115.21	45.7
■	Kyocera SN-205 Sintered Si$_3$N$_4$	2.801	3.2	.125	37.77	27.9
◇	Norton NC-350 RB-Si$_3$N$_4$ (Diamond Ground) (1976)	2.523	3.2	.212	30.87	25.5
◇	Norton NC-350 RB-Si$_3$N$_4$ (Diamond Ground) (1976)	2.396	3.2	.251	23.53	23.5
◇	Norton NC-350 RB Si$_3$N$_4$ (Diamond Ground) (1976)	2.539	3.2	.207	33.96	27.7
◀	Norton NCX-34 HP-Si$_3$N$_4$ + 8% Y$_2$O$_3$	3.372	3.391	.006	126.73	48.6
△	Raytheon Isopressed RS-Si$_3$N$_4$	2.447	3.2	.235	21.55	23.8
◢	AiResearch Slip-Cast RS-Si$_3$N$_4$ (RBN-101)	2.862	3.2	.106	37.92	32.0
◐	Kyocera SN-3 HP-Si$_3$N$_4$	3.072	3.2	.04	74.86	36.6
◐	Kyocera SN-201 Sintered Si$_3$N$_4$	3.005	3.2	.061	49.61	34.4
◆	Ford Injection Molded RS-Si$_3$N$_4$	2.742	3.2	.143	38.18	30.7
◀	Harbison-Walker HP-Si$_3$N$_4$ + 10% CeO$_2$	3.327	3.387	.018	76.78	46.9
◆	Kawecki-Berylco RS-Si$_3$N$_4$	2.509	3.2	.216	23.27	21.5
○	Kawecki-Berylco RS-Si$_3$N$_4$	2.454	3.2	.233	21.80	21.1
○	Kawecki-Berylco RS-Si$_3$N$_4$	2.349	3.2	.266	18.08	20.0
●	Indussa/Nippon Denko RS-Si$_3$N$_4$	2.084	3.2	.349	10.58	11.9
△	AiResearch Injection Molded RS-Si$_3$N$_4$ (RBN-122)	2.660	3.2	.169	32.58	30.2
⊕	GTE Laboratories Sintered Si$_3$N$_4$ + 6% Y$_2$O$_3$	3.256	3.271	.005	78.00	42.1
●	Ceradyne Ceralloy 147A, HP-Si$_3$N$_4$ + 1% MgO	3.221	3.2	.001	87.09	47.9
◐	Ceradyne Ceralloy 147-Y-1, HP-Si$_3$N$_4$ + 8% Y$_2$O$_3$	3.289	3.295	.002	33.25	45.4
◐	Ceradyne Ceralloy 147-Y, HP-Si$_3$N$_4$ + 15% Y$_2$O$_3$	3.369	3.383	.004	87.85	44.7
◆	Norton NC-350 RB-Si$_3$N$_4$ (as-fired) (1977)	2.385	3.20	.255	42.09	---

TABLE 12 (cont.)

Symbol in Figs. 14-16	Material	Bulk Density g cm⁻³	Theoret. Density[a]	P Volume Fraction Porosity	4-Point Flexure Strength ksi	Static Young's Modulus[b] 10⁶ psi
✧	Fiber Materials HP-Si₃N₄ (4% MgO)	3.171	3.216	.014	66.84	--
◇	Harbison-Walker HP-Si₃N₄ (10% CeO₂) batch 2	3.353	3.387	.010	87.89	45.0
◆	Norton NC-350 RS-Si₃N₄ (1979)	2.354	3.200	.264	37.56	22.1
⇧	Annawerk Ceranox NR-115H RS-Si₃N₄	2.571	3.200	.197	28.87	26.4
▽	AED Nitrasil RS-Si₃N₄	2.547	3.200	.204	29.06	26.3
●	AiResearch Sintered Si₃N₄ (8% Y₂O₃, 4% Al₂O₃)	3.144	3.323	.054	77.89	44.8
□	Rocketdyne SN-46 Sintered Si₃N₄ (14% Y₂O₃, 7% SiO₂)	3.433	3.47	.011	54.02	39.4
◆	Rocketdyne SN-50 Sintered Si₃N₄ (6% Y₂O₃, 4% Al₂O₃)	3.252	3.32	.020	51.60	40.8
◁	Rocketdyne SN-104 Sintered Si₃N₄ (14% Y₂O₃, 7% SiO₂)	3.365	3.47	.030	57.64	34.1
◁	Toshiba HP-Si₃N₄ (4% Y₂O₃, 3% Al₂O₃)	3.245	3.266	.006	105.7	47.5
▼	Toshiba HP-Si₃N₄ (3% Y₂O₃, 4% Al₂O₃, SiO₂)	3.204	3.261	.017	83.60	41.3
▲	Westinghouse HP-Si₃N₄, 4% Y₂O₃, SiO₂	3.250	3.247	0	90.98	44.2
◻	Georgia Tech RS-Si₃N₄	2.542	3.200	.206	26.22	23.9
◇	AME RS-Si₃N₄	2.118	3.200	.338	7.47	--
◁	AiResearch RBN-104 RS-Si₃N₄	2.785	3.200	.130	40.27	--
■	NASA/AVCO/Norton HP-Si₃N₄ (10% ZrO₂)	3.367	3.360	0	95.19	44.3
-	BuMines Sintered SiAlON (Billet 1, 5% ZrSiO₄)	2.865	--	--	26.66	28.1
-	BuMines Sintered SiAlON (Billet 2, 5% Y₂O₃)	3.049	--	--	35.33	33.0
-	BuMines Sintered SiAlON (Billet 4, 5% ZrSiO₄)	2.627	--	--	13.19	19.4
-	BuMines Sintered SiAlON (Billet 5, 5% ZrSiO₄)	2.635	--	--	17.16	22.3

[a] The theoretical density of all sintered and reaction bonded Si₃N₄ materials was assumed to be 3.200 g/cm³. The theoretical density of the HP-Si₃N₄ + 1% MgO materials was also assumed to be 3.200 g/cm³. The theoretical density of the Y₂O₃- and CeO₂-doped HPSN was computed from the nominal chemical composition assuming $\rho(CeO_2) = 7.13$ and $\rho(Y_2O_3) = 5.01$. The only exception to this is for NCX-34, which contains ~2% tungsten; the composition of NCX-34 was assumed to be 88% Si₃N₄ + 9% Y₂O₃ + 3% WC.

[b] Young's modulus determined with strain gages in 4-point flexure.

TABLE 13. SUMMARY OF ROOM-TEMPERATURE RELAXED AND DYNAMIC ELASTIC MODULUS DATA FOR VARIOUS Si_3N_4 MATERIALS

Symbol in Fig. 17	Material	Bulk Density[a] (sonic), g cm^{-3}	Theoret. Density[b]	Volume Fraction Porosity[a]	Young's Modulus (sonic)[b], 10^6 psi	Static Young's Modulus[d] and (bulk density), 10^6 psi
◆	Norton NC-132 HP-Si_3N_4 + 1% MgO	3.177	3.2	.007	45.0	48.4 (3.177)
◆	Norton NC-132 HP-Si_3N_4 + 1% MgO	3.162	3.2	.012	44.5	45.7 (3.186)
△	Norton NCX-34, HP-Si_3N_4 + 8% Y_2O_3	3.386	3.391	.001	44.5	48.6 (3.372)
△	Ceradyne Ceralloy 147-A, HP-Si_3N_4 + 1% MgO	3.312	3.2	0	45.8	47.9 (3.221)
◊	Ceradyne Ceralloy 147-Y, HP-Si_3N_4 + 15% Y_2O_3	3.370	3.333	.004	41.6	44.7 (3.369)
◊	Ceradyne Ceralloy 147-Y-1, HP-Si_3N_4 + 8% Y_2O_3	3.218	3.295	.023	42.8	45.4 (3.289)
◊	Kyocera SN-3 HP-Si_3N_4	3.060	3.2	.044	35.9	36.6 (3.072)
◁	Kyocera SN-201 Sintered Si_3N_4	2.994	3.2	.064	34.7	34.4 (3.005)
◐	Kyocera SN-205 Sintered Si_3N_4	2.819	3.2	.119	27.6	27.9 (2.801)
◑	Norton NC-350 RS-Si_3N_4 (1976)	2.470	3.2	.228	25.6	25.5 (2.523)
⊕	Norton NC-350 RS-Si_3N_4 (1976)	2.392	3.2	.253	22.7	23.5 (2.396)
◒	Norton NC-350 RS-Si_3N_4 (1976)	2.524	3.2	.211	26.8	27.7 (2.539)
●	Kawecki-Berylco (KBI) RS-Si_3N_4	2.325	3.2	.273	18.6	21.5 (2.509)
◖	KBI RS-Si_3N_4	2.347	3.2	.267	18.8	21.1 (2.454)
◗	KBI RS-Si_3N_4	2.432	3.2	.240	23.1	20.0 (2.349)
○	Ford (IM) RS-Si_3N_4	2.762	3.2	.137	28.3	30.7 (2.742)
▷	AiResearch (SC) RS-Si_3N_4	2.881	3.2	.099	31.4	32.0 (2.862)
▽	AiResearch (IM) RS-Si_3N_4	2.646	3.2	.173	26.6	30.2 (2.660)
□	Raytheon (IP) RS-Si_3N_4	2.402	3.2	.249	21.5	23.8 (2.447)
✕	Fiber Materials HP-Si_3N_4 (4% MgO)	3.197	3.216	.006	46.2	--
◆	Harbison-Walker HP-Si_3N_4 (10% CeO_2) batch 2	3.343	3.387	.014	44.6	45.0 (3.353)
◐	Norton NC-350 RS-Si_3N_4 (1979)	2.422	3.200	.243	22.7	22.1 (2.354)
◇	Annawerk Ceranox NR-115H RS-Si_3N_4	--	3.200	--	--	26.4 (2.571)
▶	AED Nitrasil RS-Si_3N_4	2.561	3.200	.200	25.7	26.3 (2.547)

TABLE 13 (cont.)

Symbol in Fig. 17	Material	Bulk Density (sonic) g/cm³ [a]	Theo. Density [b]	Volume Fraction Porosity [a]	Young's Modulus (sonic) [c] 10⁶ psi	Static Young's Modulus [d] (bulk density), 10⁶ psi
■	Toshiba HP-Si₃N₄ (4% Y₂O₃, 3% Al₂O₃)	3.253	3.266	.004	47.0	47.5 (3.245)
◼	Toshiba HP-Si₃N₄ (3% Y₂O₃, 4% Al₂O₃, SiO₂)	3.217	3.261	.013	42.0	41.3 (3.204)
▲	Westinghouse HP-Si₃N₄ (4% Y₂O₃, SiO₂)	3.286	3.247	0	44.6	44.2 (3.250)
●	Georgia Tech RS-Si₃N₄	2.590	3.200	.191	25.6	23.9 (2.542)
◼	NASA/AVCO/Norton HP-Si₃N₄ (10% ZrO₂)	3.338	3.360	.007	45.0	44.3 (3.367)
—	BuMines Sintered SiAlON (Billet 1, 5% ZrSiO₄)	2.813	—	—	30.6	28.1 (2.865)
—	BuMines Sintered SiAlON (Billet 2, 5% Y₂O₃)	3.049	—	—	34.5	33.0 (3.049)
—	BuMines Sintered SiAlON (Billet 4, 5% ZrSiO₄)	2.601	—	—	19.2	19.4 (2.627)
—	BuMines Sintered SiAlON (Billet 5, 5% ZrSiO₄)	2.634	—	—	22.0	22.3 (2.635)

[a] Average values for dynamic modulus samples
[b] Theoretical density computed from nominal chemical composition.
[c] Dynamic elastic modulus determined by flexural resonant frequency method.
[d] Determined with strain gage transducers during 4-point flexure strength test at 0.02 ipm.

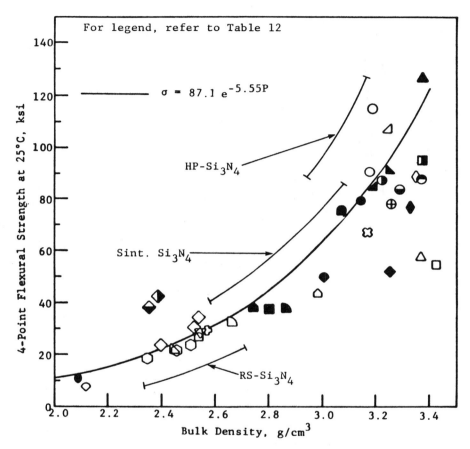

Figure 14. Room-temperature flexural strength vs. bulk density for various Si_3N_4 materials.

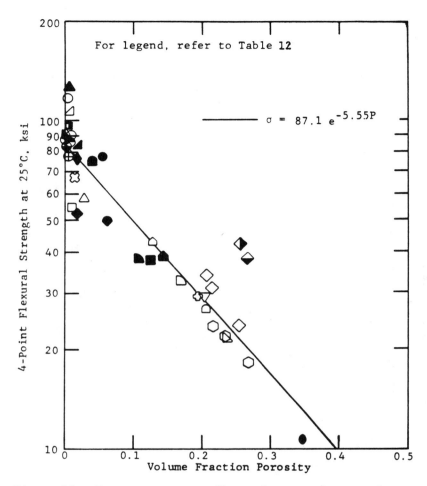

Figure 15. Room-temperature flexural strength vs. volume fraction porosity for various Si_3N_4 materials.

it is for the lower density sintered and reaction-bonded materials. That is, the critical strength-controlling flaws in the lower density materials are expected to be pores or pore clusters, while fracture sources in high density hot-pressed materials have typically been traced to large grains, metal inclusions, unreacted particles, surface or subsurface machining damage, or a single isolated pore not affecting the overall density.

The hot-pressed materials with strengths greater than 100 ksi are Norton NC-132 (1% MgO), Norton NCX-34 (8% Y_2O_3), and Toshiba (4% Y_2O_3, 3% Al_2O_3) materials. For RS-Si_3N_4, it is noted that the strength of two Norton NC-350 materials is ~40 ksi, well above the trend line.

Fracture surface analysis was conducted on this program with a 6-60X stereoscopic optical microscope with photographic attachment. The objective of this examination was to attempt to identify the general nature of the fracture origins. Room temperature fracture sources in Si_3N_4 are tabulated in Table 11. It is observed that the Norton hot-pressed materials (NC-132 and NCX-34) were relatively processing-mature in that the fracture sources that could be identified were predominantly machining induced flaws at the diamond-ground surface. All other hot-pressed materials exhibited fracture origins consisting of inclusions and other subsurface processing defects.

The fracture sources in the various sintered Si_3N_4 materials were a mixture of inclusions and porosity (surface and subsurface). Porosity and inclusion particles were also the predominant fracture sources in RS-Si_3N_4 materials. The inclusions are most probably unreacted silicon particles and iron-rich particles. Iron is a common impurity (and sintering aid) in RS-Si_3N_4. Spectrographic analysis results (Table 7) indicate as much as 1.3% Fe impurity in these materials. However, porosity is the predominant microstructural feature in RS-Si_3N_4 that can be described as detrimental to those properties necessary for high performance.

Evans[37] presents an enlightening analysis of the sensitivity of HP-Si_3N_4 materials to various types of processing defects. As summarized in Figure 18, the fracture stress is relatively insensitive to tungsten carbide inclusions, and progressively more sensitive to Fe, C, and voids. Unreacted silicon particles and surface cracks were found to have the strongest effect on strength.

The relaxed elastic moduli obtained during flexure testing of all Si_3N_4 materials are plotted as a function of porosity in Figure 16 (data compiled in Table 12). The dynamic moduli (obtained by a flexural resonant frequency method) of all Si_3N_4 materials are compiled in Table 13 (where the static moduli are repeated for comparison). The 25°C relaxed and dynamic elastic moduli for all Si_3N_4 materials are plotted as a function of volume fraction porosity in Figure 17. Good agreement is obtained between the static and dynamic moduli at room temperature. The data fit for all Si_3N_4 static and dynamic moduli has the exponential form:

$$E = 45.1e^{-2.99P}$$

6.2 SiC MATERIALS

Silicon carbide materials are processed in hot-pressed, sintered, and silicon-densified forms. All are essentially fully dense. There is no analogy to porous reaction-sintered silicon nitride for the SiC system.

Silicon carbides have high elastic modulus compared to Si_3N_4 materials. The Young's modulus is typically 50-60 x 10^6 psi. This is ~50% higher than the elastic modulus of dense HP-Si_3N_4. The strength of hot-pressed and sintered SiC is intermediate to the strength of hot-pressed and reaction sintered Si_3N_4. Room-temperature strength in SiC is usually controlled by the grain size. Hot-pressed SiC using Al_2O_3 as a densification aid has higher strength than when B_4C is used. Al_2O_3 inhibits grain growth; the grain size is typically 1-2 μm, and the strength is

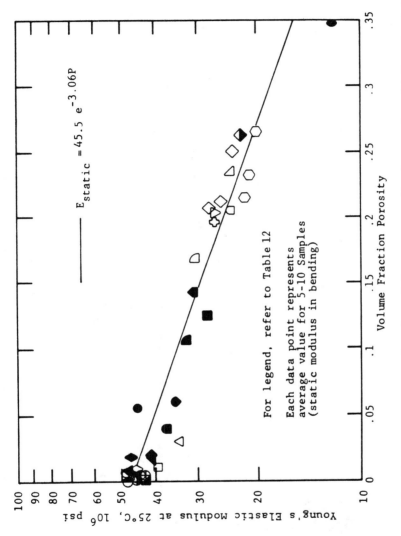

Figure 16. Room-temperature elastic modulus vs. porosity for various Si_3N_4 materials.

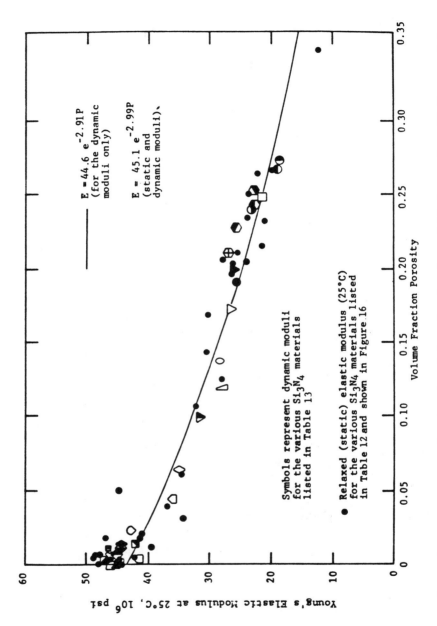

Figure 17. Comparison of relaxed and dynamic elastic moduli for various Si_3N_4 materials.

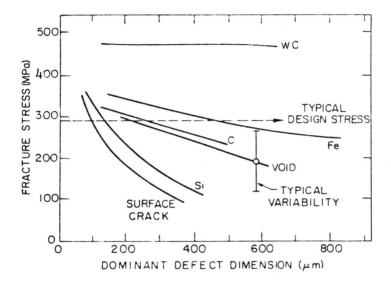

Figure 18. Strength, size relations for various fracture-initiating defects in silicon nitride (from Evans[37]).

high. The converse is true for HP-SiC doped with B_4C, where the grain size ranges from 10-40 μm.

Sintered SiC is processed in α- and β-SiC crystal structures (predominantly hexagonal and cubic, respectively). Bend strengths are typically 45-65 ksi. The β-SiC structure appears consistently stronger than α-SiC. Fracture origins in α-SiC are often large (~50 μm) individual needles/platelets that have experienced exaggerated grain growth during processing. Single isolated surface-connected pores are also found in α-SiC. Both forms of SiC fracture predominantly in the transgranular mode. It is for this reason that fracture phenomena in SiC are more often associated with grain size, rather than porosity or intergranular phases. The predominance of transgranular fracture also makes fracture surface analysis and the identification of fracture origins more difficult than in $HP-Si_3N_4$, where the fracture path is usually intergranular. Details are presented in the following paragraphs.

6.2.1 Strength-Grain Size Relation

The results of the microstructural analysis of all SiC materials evaluated to date are summarized in Table 14. In this tabulation we include the density, mean room-temperature fracture strength, average and maximum grain size, and the location of any porosity. The approximate grain size determination was made by an overall visual inspection of the micrographs, and not quantitatively determined by lineal analysis or other statistical techniques. In assessing the location of porosity, void features with rounded corners were judged to be actual porosity, whereas void features with sharp/angular corners were judged to be polishing pullouts. The comments in Table 14 regarding fracture origins and fracture mode were derived from the fracture surface analysis of bend bars broken at room temperature. The strength and elastic moduli of all SiC materials evaluated, as correlated with density/porosity, are tabulated in Tables 15 and 16. The modulus-porosity data for SiC are plotted in Figure 19.

TABLE 14. MICROSTRUCTURAL ANALYSIS SUMMARY FOR SiC MATERIALS

Material	Density g cm⁻³	Mean 25°C Bend Strength, psi	Approximate Grain Size, μm Avg.	Approximate Grain Size, μm Max.	Location of Porosity[c]	Room Temperature Fracture Origins; General Fracture Mode
HOT-PRESSED SiC						
Ceradyne Ceralloy 146-A (2% Al$_2$O$_3$)	3.22	60,060	4-12	30	In grain boundaries	Undetermined; primarily transgranular
Ceradyne Ceralloy 146-I (2% B$_4$C)	3.21	45,620	10-30	50	∼3/4 in grain boundaries ∼1/4 within grains	Some machining damage, primarily undetermined; transgranular
Norton NC-203 (∼2% Al$_2$O$_3$)	3.32	101,810	1-4	10	In grain boundaries	Undetermined; transgranular
SINTERED SiC						
General Electric β-SiC	3.04	63,770	0.5-2(a)	100(a)	Undetermined	Primarily inclusions or large grains; transgranular
Carborundum 1977 α-SiC (SC)	3.16	44,230	2-8	15	∼2/3 in grain boundaries ∼1/3 within grains	Primarily inclusions or large grains; transgranular
Kyocera 1980 SC-201 α-SiC	3.14	56,080	1.5-5	10	In grain boundaries	Primarily undetermined, some surface and subsurface pores; transgranular
Carborundum 1981 α-SiC (IM)	3.09	47,230	2-5	15-18	Mostly intergranular	Primarily surface and subsurface 1-4 μm pores
ESK α-SiC	3.16	41,660	3-8	60-120	∼2/3 in grain boundaries ∼1/3 within grains	Surface and subsurface porosity.
SILICONIZED SiC						
Norton NC-435	2.96	57,200	1-6	12	In grain boundaries	Pores or large grains, some undetermined; intergranular
UKAEA/BNF Refel[a]						
• As-processed	3.09	33,600	0.5-5	15	In grain boundaries	Some large grains, primarily undetermined, combination interranular and transgranular
• Diamond-ground	3.11	44,920	0.5-3	15-50	In grain boundaries	Large grains; combination intergranular and transgranular
Norton NC-430	3.1	30,420	2-10[d]	50-175[c]	Mostly between small grain boundaries, some within large grains	Undetermined; combination intergranular and transgranular
Coors (1979, SC-1)	3.0	50,620	1.5-6	12	In grain boundaries, between Si phase	Mostly undetermined, some machining damage
Coors (1981, SC-2)	3.09	45,410	0.5-3.5 5-12	20	—	Slightly less Si in Coors 1981.
Coors (1982, SC-2)	3.10	43,310	0.5-2 5-10	15	—	Primarily large grains.

[a]Diamond-ground and as-processed refer to the material condition for the strength tests, the microstructure analyses for grain size and Si distribution were made on bulk material well below the surface. The near-surface microstructural features are different for the diamond-ground and as-processed conditions.
[b]From overall visual inspection of micrographs, but statistically determined by lineal analysis, etc.
[c]Void features with rounded corners were judged to be porosity, void features with sharp/angular corners were judged to be polishing pullouts.
[d]Bimodal size distribution of SiC grains.

TABLE 15. SUMMARY OF ROOM-TEMPERATURE STRENGTH AND POROSITY DATA FOR VARIOUS SiC MATERIALS

Material	Bulk Density, g cm^{-3}	Theoret. Density[a]	Volume Fraction Porosity[b]	4-Point Bend Strength, ksi
Norton NC-435 Siliconized SiC, batch 1	2.936	3.0396	.0341	50.5
Norton NC-435 Siliconized SiC, batch 3	2.997	3.0396	.0140	66.0
Norton NC-435 Siliconized SiC, batch 4	2.962	3.0396	.0255	55.2
General Electric Sintered β-SiC	3.032	3.217	.0575	63.8
Carborundum Sintered α-SiC (1977)	3.159	3.217	.0180	44.2
UKAEA/BNF Refel Si/SiC, Diamond Ground	3.101	3.1283	.0087	44.9
UKAEA/BNF Refel Si/SiC, As-Processed	3.090	3.1283	.0122	33.6
Ceradyne Ceralloy 146A, HP-SiC (2% Al$_2$O$_3$)	3.224	3.2293	.0016	60.1
Ceradyne Ceralloy 146I, HP-SiC (2% B$_4$C)	3.211	3.1993	0	45.6
Norton NC-203 HP-SiC (2% Al$_2$O$_3$)	3.318	3.375	.017	101.8
Coors Si/SiC (1979, SC-1)	3.001	3.099	.032	50.6
Kyocera SC-201 Sintered SiC	3.148	3.217	.021	56.1
Norton NC-430 Si/SiC	3.105	3.217	.035	30.4
Carborundum 1981 Hexoloy SX-05 SSC	3.117	3.217	.031	47.2
Coors Si/SiC (1981, SC-2)	3.097	3.099	0	45.4
Coors Si/SiC (1982, SC-2)	3.100	3.099	0	43.3
General Electric Silcomp Si/SiC (CC)	2.886	3.062[c]	.057	26.0
ESK Sintered α-SiC	3.160	3.217	.018	41.7

[a]Theoretical density computed from nominal chemical composition assuming ρ_{SiC} = 3.217, ρ_{Si} = 2.33, ρ_{B_4C} = 2.52, $\rho_{Al_2O_3}$ = 3.97 g cm^{-3}.

The effective theoretical density for NC-435 and Refel siliconized SiC was computed by relation $\rho = v_1\rho_1 + v_2\rho_2$, assuming 20 vol% and 10 vol% silicon phase, respectively.

The effective theoretical density of all other materials was computed from their nominal chemical composition using the relation $1/\rho = x_1/\rho_1 + x_2/\rho_2$, where x_1 and x_2 refer to the wt% of each phase present.

[b]$P = 1 - \rho_{Bulk}/\rho_{Theoret}$.

[c]Estimate.

TABLE 16. SUMMARY OF ROOM-TEMPERATURE RELAXED AND DYNAMIC ELASTIC MODULUS DATA FOR VARIOUS SiC MATERIALS

Symbol in Fig. 19 Static Modulus	Symbol in Fig. 19 Dynamic Modulus	Material	Bulk Density (sonic)[a] g cm^{-3}	Theoret. Density[b]	Volume Fraction Porosity	Young's Modulus (sonic)[c], 10^6 psi	Static Young's Modulus[d] (with corresponding density), 10^6 psi
●	○	Norton NC-435 Siliconized SiC, batch 1	2.953	3.217	.0821	49.1	53.9 (2.936)
■	□	Norton NC-435 Siliconized SiC, batch 3	2.973	3.217	.0758	49.6	49.5 (2.997)
◆	◇	Norton NC-435 Siliconized SiC, batch 4	2.991	3.217	.0703	50.6	48.8 (2.962)
▲	△	General Electric Sintered β-SiC	3.031	3.217	.0578	54.6	54.8 (3.032)
▲	△	Carborundum Sintered α-SiC (1977)	3.159	3.217	.0180	62.1	58.2 (3.159)
▲	△	UKAEA/BNF Refel Si/SiC, Diamond Ground	3.086	3.217	.0407	56.2	57.5 (3.101)
▲	△	UKAEA/BNF Refel Si/SiC, As-Processed	3.049	3.217	.0522	49.1	52.5 (3.090)
●	○	Ceradyne Ceralloy 146A, HP-SiC + 2% Al$_2$O$_3$	3.225	3.229	0	63.5	67.2 (3.224)
●	○	Ceradyne Ceralloy 146I, HP-SiC + 2% B$_4$C	3.229	3.217	0	62.6	65.3 (3.211)
◆	◇	Norton NC-203 HP-SiC (2% Al$_2$O$_3$)	3.331	3.375	.013	64.7	--
●	○	Norton NC-430 Si/SiC	3.130	3.217	.027	58.4	58.7 (3.105)
▲	△	Coors Si/SiC (1979, SC-1)	3.026	3.099	.024	52.3	51.1 (3.001)
●	○	Kyocera SC-201 Sintered SiC	3.113	3.217	.032	58.5	60.5 (3.148)
⊖	—	Carborundum 1981 Hexoloy SX-05 SSC	--	3.217	--	--	55.9 (3.117)
■	—	Coors Si/SiC (1981, SC-2)	3.118	3.099	0	58.0	53.8 (3.097)
◆	—	Coors Si/SiC (1982, SC-2)	--	3.099	--	--	54.2 (3.100)
▲	—	General Electric Silcomp Si/SiC (CC)	2.919	3.062[e]	.047	48.0	43.7 (2.886)
⊖	—	ESK Sintered α-SiC	--	3.217	--	--	61.0 (3.160)

[a] Average values for dynamic modulus samples.
[b] Theoretical density computed from nominal chemical composition only for Ceradyne 146A. All other materials: Theoretical density assumed to be ρ_{Th} = 3.217 g cm^{-1}.
[c] Dynamic elastic modulus determined by flexural resonant frequency method.
[d] Determined with strain gage transducers during 4-point flexure strength test (0.02 ipm crosshead deflection).
[e] Estimate.

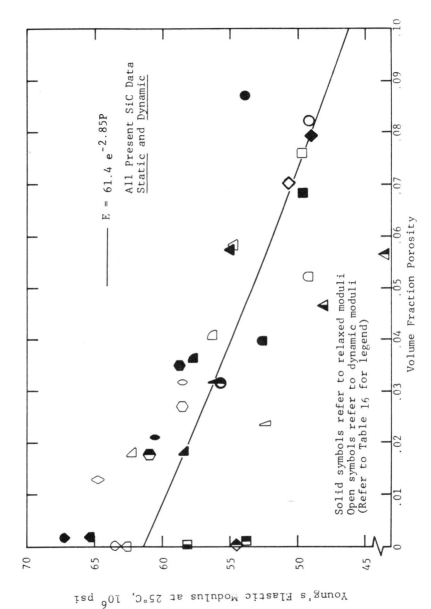

Figure 19. Comparison of relaxed and dynamic elastic moduli for various SiC materials (including hot-pressed, sintered, and silicon-densified forms of SiC).

(a) <u>Hot-pressed SiC</u>. The room-temperature strength of SiC can be correlated with grain size in general, as would be expected since the fracture path in SiC is predominantly transgranular. For instance, the strengths of the three fully dense hot-pressed (HP) SiC materials evaluated on this program are compared in Table 14.* The reflected light micrographs of these three materials are presented in Figure 20. Norton NC-203 HP-SiC, doped with ~2% Al_2O_3, has very high strength (σ >100 ksi) since it is so fine-grained. Its grain size averages 1-4 μm, with the largest grains being ~10 μm in size (see Table 14). All grains in NC-203 are equiaxial. This information is seen in the micrograph presented in Figure 20c. The strength of the companion Ceradyne material, Ceralloy 146-A which also contains nominally 2% Al_2O_3 as a densification additive, is much lower, i.e., ~60 ksi at room temperature. Figure 20a illustrates a larger grain size for Ceralloy 146-A than for NC-203, i.e., and average grain size ~4-12 μm, with a ~30 μm maximum. Thus, the grain size of the Ceradyne material is a factor of four larger than the grain size of NC-203, and the strength of the Ceralloy 146-A is roughly half that of NC-203. This is exactly what would be predicted from the known strength-grain size relation for ceramics; that is, the strength being inversely proportional to the square root of the grain diameter.[38,39]

The room-temperature strength of Ceradyne Ceralloy 146-I HP-SiC, which is doped with ~2% B_4C to promote densification, is even lower, ~45 ksi as shown in Table 14. Figure 20b illustrates the microstructure of Ceralloy 146-I. The average grain size is shown to be even larger than for the Al_2O_3-doped HP-SiC materials (i.e., 10-30 μm average grain diameter, with a maximum of ~50

---

*In this discussion of grain size we are only concerned with room temperature strength; at elevated temperature other factors affect the strength, mainly the amount of oxide intergranular phase and resulting subcritical crack growth (refer to Section 7.2).

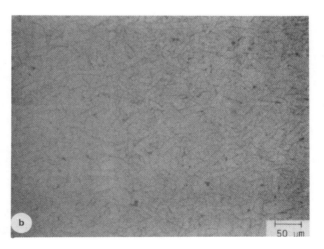

Figure 20. Reflected light micrographs of hot-pressed silicon carbide materials.

(a) Ceradyne Ceralloy 146-A HP-SiC (2% Al_2O_3) etched 9.6 min with boiling Murikami's reagent.

(b) Ceradyne Ceralloy 146-I HP-SiC (2% B_4C) etched 8.6 min with boiling Murikami's reagent.

Figure 20 (cont.)

(c) Norton NC-203 HP-SiC ($\sim$2% Al_2O_3) etched 9.6 min with boiling Murikami's reagent.

μm). Again, the data follow the inverse square root strength-grain diameter relation. The grain size of Ceralloy 146-I is about an order of magnitude larger than that of NC-203, and the room-temperature strength is about one third that of NC-203. The reason for the factor of three grain size difference in the two Ceradyne materials has to do with the effect of the oxide additive on grain growth during processing. The presence of Al_2O_3 inhibits grain growth; B_4C promotes grain growth in SiC. The resulting effect on strength is shown in the present data. The grain growth in the B_4C-doped material is apparently anisotropic, since the grain morphology in Ceralloy 146-I is largely tabular, with an aspect ratio of approximately 3:1.

(b) <u>Sintered SiC</u>. Sintered SiC materials from General Electric, Carborundum, Kyocera, and ESK[*] were evaluated on this program. Sintered SiC is processed in the α and β crystal structures (predominantly hexagonal and cubic, respectively). The General Electric material used to be referred to as boron-doped β-SiC. The materials from Carborundum (1977 and 1981 vintages) and Kyocera (SC-201, 1980 vintage) are of the α crystalline form. The ESK material is also α-SiC. All, except ESK, employ boron as the primary sintering aid. Emission spectrographic analysis indicates all have ~0.4-0.5 wt% boron and ~0.2-0.3 wt% iron impurities (refer to Table 7). On the other hand, Al_2O_3 was the primary sintering aid for the ESK material (high Al impurity as shown in Table 7).

Strength data for these sintered SiC materials are shown in Table 14. The reflected light micrographs for these materials are presented in Figure 21. The various microstructural analysis parameters are summarized in Table 14. The β-SiC structure appears consistently stronger than α-SiC. Table 14 shows that General Electric β-SiC is a nominally 65 ksi material. Figure

---

[*]Elektroschmelzwerk Kempten, Kempten, FRG.

Materials Evaluation 87

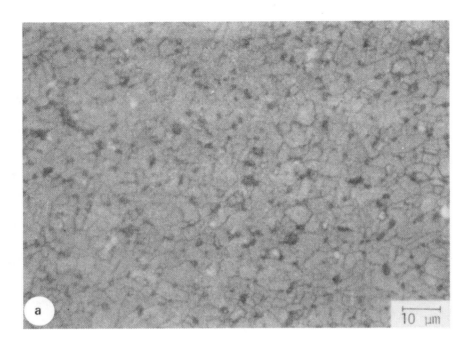

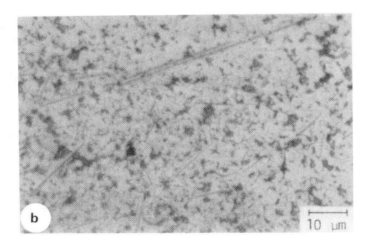

Figure 21. Reflected light micrographs of sintered silicon carbide materials.

(a) 1980 Kyocera SC-201 sintered α-SiC etched 8 min with boiling Murikami's reagent.

(b) General Electric sintered β-SiC etched 5 min with boiling Murikami's reagent (α-etch).

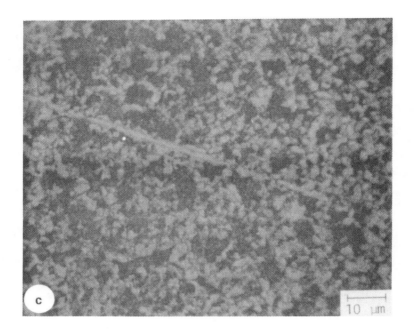

Figure 21 (cont.)

(c) General Electric sintered β-SiC etched 5 min with boiling Murikami's reagent (α-etch) and 1.5 min with a fused salt mixture of KOH and KNO_3 (β-etch).

(d) 1977 Carborundum sintered α-SiC etched 10.3 min with boiling Murikami's reagent.

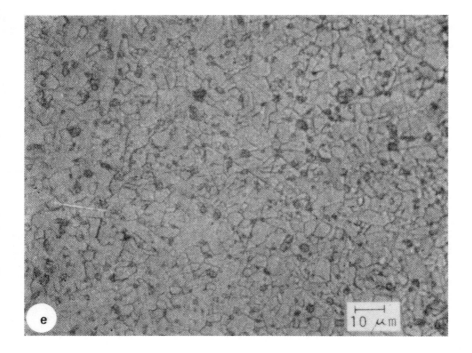

Figure 21 (cont.).

(e) 1981 Carborundum sintered α-SiC polished and etched 7.5 min with boiling Murikami's reagent.

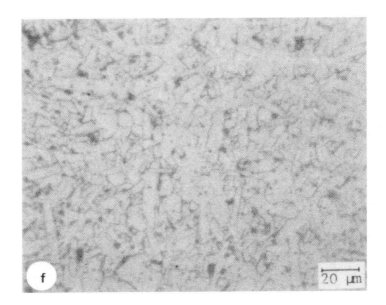

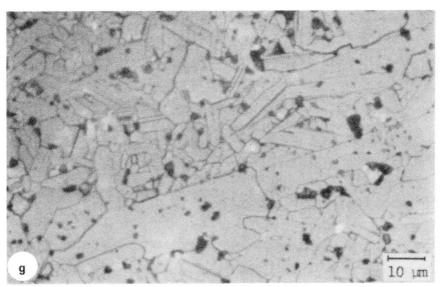

Figure 21 (cont.)

(f) ESK sintered α-SiC polished and etched 8 min with boiling Murikami's reagent.

(g) ESK sintered α-SiC.

21c illustrates that this high strength is achieved by obtaining a very fine β-SiC grain size (typically 0.5-2 μm). This material actually has a small amount of the α-structure present. Figure 21b shows some evidence of long needles/platelets of α-SiC in the overall structure. The X-ray diffraction results shown in Table 6 confirm that a small amount of the α-structure can be present in this β-SiC from General Electric. The acicular α-phase grains can become quite large (i.e., ~100 μm long in Figure 21c), but apparently are so few in number as to not override the influence of the much smaller β-crystals and not be detrimental to the strength of the overall structure. Since this would appear to be a violation of the weakest link theory, perhaps these long α-needles impart a whisker-like reinforcement to the structure. The two etching procedures used to delineate the features of the α and β crystal phases shown in Figures 21b and 21c are explained in Table 9. The development of the microstructure in General Electric β-SiC is described in detail by Johnson and Prochazka.[40]

The α-SiC materials from Carborundum and Kyocera are lower in strength compared to General Electric β-SiC. Carborundum 1977 α-SiC has a room-temperature 4-point bend strength of nominally 45 ksi, whereas the strength of Kyocera 1980-vintage SC-201 α-SiC is on the order of 55 ksi. The reflected light micrographs shown in Figures 21a and 21d illustrate the equiaxed grain configuration for these materials. The grain diameter of the Kyocera α-SiC is slightly smaller than that of the Carborundum material. The average grain diameter of Kyocera SC-201 ranges from ~1.5 to 5 μm. An occasional 10 μm grain is observed. The average grain size range and maximum grain diameter of 1977 Carborundum α-SiC are 2-8 μm and 15 μm, respectively. Therefore, the higher strength of the Kyocera material appears to be related to its slightly smaller grain size. No evidence of exaggerated growth of α grains is observed for these materials.

Carborundum 1981 sintered α-SiC (Hexoloy SX-05) was injection-molded and supplied in the as-fired condition (the 1977 material was slip-cast and diamond ground). The 1981 material had slightly lower density. Both materials had similar phase content (α-SiC, with similar polytype structures) and cation impurities. The microstructures of these materials are similar. The more recent material was slightly stronger (refer to Table 14).

The ESK material exhibited a room-temperature 4-point bend strength of 41.7 ksi. This is on the low end of the range of strengths exhibited by the other sintered SiC materials evaluated. Microstructural analysis yields two explanations: exaggerated α-grain growth and large intergranular porosity, as illustrated in the micrographs shown in Figure 21 f and g. The average size of the α-grains is 3-8 μm. Most of these grains are not equiaxed, but have an aspect ratio of ~1:7. A number of large α-grains have grown to platelet size, 60-120 μm. Some pores are as large as 10 μm in diameter. Fracture surface analysis indicated that surface and subsurface pores were the primary fracture origins at room temperature, so it may be that the large α-grains had no effect on the strength. At this time the composition of the light areas on the micrograph shown in Figure 21g is not known; they may be Al_2O_3-rich areas.

(c) <u>Silicon-Densified SiC</u>. Bend strength data for the silicon-densified SiC materials are presented in Table 14. The micrographs are shown in Figure 22, and the microstructural parameters are also summarized in Table 14. The strongest materials at 25°C are Norton NC-435 and Coors 1979 SC-1 Si/SiC (51-57 ksi). The micrographs illustrate that these materials have the smallest grain size, which accounts for their high strength. The average grain size range for these two materials is ~1-6 μm; the maximum grain size observed is ~12 μm. The reflected light micrographs in Figures 22a and 22f illustrate that the grains of NC-435 are equiaxial, whereas about 25% of the Coors 1979 SC-1 Si/SiC grains are tabular with approximate aspect ratio of 10:1. The distribution of the continuous silicon phase is about the same in both materials, uniform, and appearing to be ~10-20% by volume in extent.

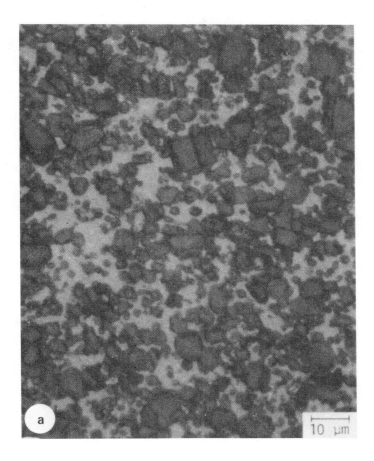

Figure 22. Reflected light micrographs of siliconized silicon carbide materials.

(a) Norton NC-435 siliconized SiC electrolytically etched 4 min with 20% KOH.

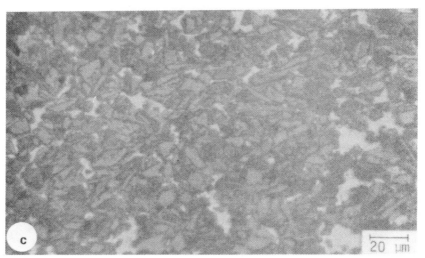

Figure 22 (cont.)

(b) UKAEA/British Nuclear Fuels Refel siliconized SiC; microstructure of diamond-ground material, electrolytically etched 2 min with 20% KOH.

(c) UKAEA/British Nuclear Fuels Refel siliconized SiC; dense interior microstructure of as-processed material, electrolytically etched 3 min with 20% KOH.

Materials Evaluation 95

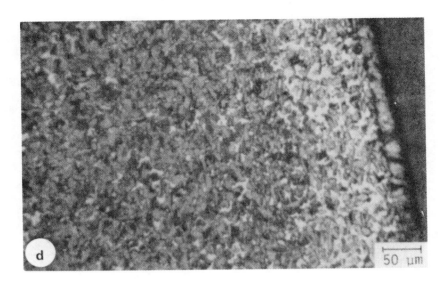

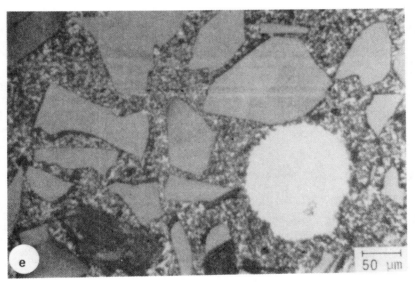

Figure 22 (cont.)

(d) UKAEA/British Nuclear Fuels Refel siliconized SiC, as-processed with high purity SiC and silicon-rich surface layers intact; electrolytically etched 3 min with 20% KOH.

(e) Norton NC-430 siliconized SiC, electrolytically etched 2 min with 20% KOH.

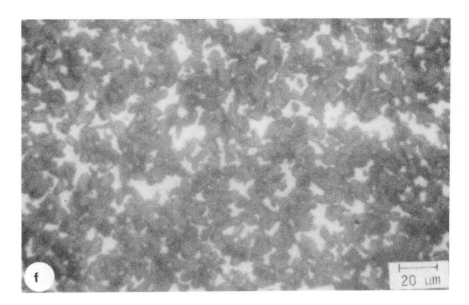

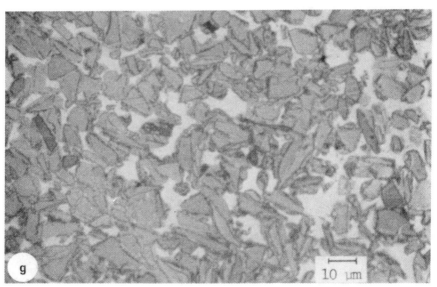

Figure 22 (cont.)

(f) Reflected light micrograph of 1979 Coors SC-1 siliconized SiC, electrolytically etched 3.0 min with 20% KOH.

(g) Reflected light micrograph of 1981 Coors SC-2 siliconized SiC, electrolytically etched 2.5 min with 20% KOH.

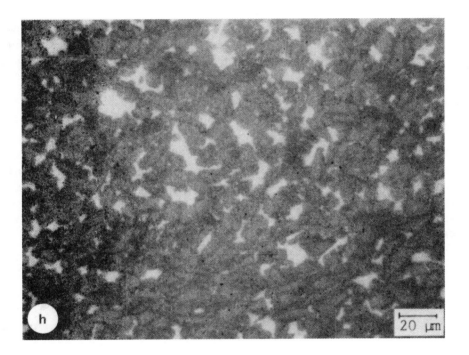

Figure 22 (cont.)

(h) **Reflected** light micrograph of 1982 Coors SC-2 **siliconized** SiC, electrolytically etched 1.5 min with 20% KOH.

The UKAEA/BNF Refel Si/SiC was received from the manufacturer in two conditions, diamond-ground and as-processed. The difference in these two conditions mainly involves the near-surface microstructure, and was described by Kennedy et al.,[41] and discussed in a previous report on this program.[42] With reference to Figure 22d, as-processed material has a highly crystalline ~25 μm layer of silicon carbide on the outer surface. Below this thin surface layer is a ~100 μm thick layer of silicon-rich material (~38 vol% silicon by manufacturer's estimate). Both of these outer layers are clearly seen in Figure 22d. Below these two outer layers of the as-processed material is the dense bulk Si/SiC, which the manufacturer states to contain 8-10 vol% continuous silicon phase. The micrograph presented in Figure 22d clearly shows all three microstructural regions of the as-processed material. It would be expected that the diamond-ground version of Refel would have two outer layers removed and exhibit the microstructure of the dense bulk material. This fact is confirmed in the micrographs in Figures 22b and 22c, and summarized in Table 14: the interior sections of both diamond-ground and as-processed material had average grain diameters ranging from approximately 0.5 to 4 μm, and the silicon phase looked relatively uniformly distributed in both.

Another silicon-densified SiC material evaluated on this program is Norton NC-430. The strength is seen in Table 14 to be only 30 ksi, the lowest of the Si/SiC materials. The microstructure shown in Figure 22e explains this; NC-430 is an extremely coarse-grained material. Actually, NC-430 exhibits a bimodal distribution of SiC grains: a network of relatively fine 2-10 μm SiC grains, and a distribution of extremely angular, extremely large 50-175 μm SiC grains. This structure is bonded together by the continuous silicon metal phase, but it is rather nonuniform in NC-430, as evidenced by the large silicon-rich region shown in the micrograph presented in Figure 22e.

It is noted that Coors has supplied three Si/SiC materials to this program. Spectrographic analysis results reveal about the same level of cation impurities in all the materials (Table 7). X-ray studies indicate all generations of material contain similar SiC polytype structures (Table 6). The most recent material, 1982 SC-2, has about the same density (3.097 g cm^{-3}) as 1981 SC-2, and about 3% higher than 1979 SC-1.

Reflected light micrographs of these three materials are presented in Figures 22f-h. The micrographs reveal that 1982 SC-2 has about the same silicon metal content as 1981 SC-2. The earlier 1979 SC-1 material had slightly higher silicon content, and a smaller and more uniform SiC grain size. Again, like 1981 SC-2, there are two distinct size ranges of SiC grains in 1982 SC-2 Si/SiC. In the more recent material the finer, submicron grains appear to be evenly distributed around the larger grains, instead of clustered together as was seen in the 1981 material.

6.2.2 Effect of Porosity

All forms of SiC thus far evaluated on this program (i.e., hot-pressed, sintered, and silicon-densified) are essentially nearly fully dense, at least compared to the range of densities obtained for the various processing methods for silicon nitride (i.e., fully dense hot-pressed to reaction-sintered, which has up to 20% porosity). Therefore, porosity is not a major limiting factor in the behavior of SiC, as it is for RS-Si_3N_4, for example. High elastic modulus and high thermal expansion are usually cited as the most limiting factors of SiC from a properties standpoint.

However, porosity in SiC does exist and can affect properties. Thus it is informative to view the porosity in SiC. The micrographs shown for hot-pressed (Figure 20), sintered (Figure 21), and silicon-densified (Figure 22) forms of SiC show that the porosity is almost entirely in the grain boundaries. Knowing this location of the porosity in SiC helps to interpret the transgranular fracture mode observed for all forms of SiC (see Table 14). Rice[38] points out that in general, other factors being equal, the advancing crack front usually follows the most

porous path. For example, intergranular failure occurs with
predominantly intergranular pores; and conversely, a transgranular fracture mode is obtained with intragranular porosity. The
predominance of a transgranular fracture mode in SiC in the presence of intergranular porosity means that either (a) there is not
enough grain boundary porosity in SiC to make the fracture path
intergranular, or (b) the low cleavage energy of the SiC crystals
promotes transgranular fracture and overrides the effect of the
intergranular porosity. We suspect that the latter explanation
may be the reason.

6.2.3 Fracture Origins

The fracture origins for all forms of SiC evaluated on this
program are summarized in Table 14. Strength is correlated (inversely) with grain size in SiC since the fracture mode is transgranular, and since minimal porosity and intergranular impurity
phases are present. Fracture origins in SiC are typically large
grains, with some impurity inclusions and some isolated surface-connected porosity. Fracture in both α and β forms of sintered
SiC can sometimes be traced to exaggerated grain growth of the α-phase, leading to very acicular (needle-shaped) SiC crystals.
However, sometimes this microstructural feature is there, but not
found to be detrimental. While a noticeable number of large
grains/α-needles were observed in the 1981 Carborundum sintered
α-SiC, the fracture origins were mostly surface and subsurface
pores. The most dramatic example of this is the ESK material,
for which the strength-limiting flaw was almost always porosity,
even though some α-platelets were 60-120 μm in size.

In general, fracture source identification in SiC is more
difficult than it is in Si_3N_4. This is due to the transgranular
fracture path in all forms of SiC, and to the heterogeneous nature of the siliconized-SiC materials. In such cases the conditions are not favorable for the development of fracture surface
features that permit the straightforward assessment of the source
of fracture.

7. Elevated Temperature Strength and Time Dependence

The most fundamental driving force for the use of structural ceramics in heat engine applications is the ability to extend operating temperatures upward beyond the limits of metallic superalloys. Therefore, elevated temperature properties for ceramics are generated, thereby exposing deficiencies that are often related to processing variables. This leads to process iterations with the aim of improving elevated temperature performance.

Perhaps the most crucial issue in the high temperature behavior of ceramics is time dependence of strength. This is the central theme in the developing of life prediction methodologies. The most general term to describe the time dependence of strength in silicon base ceramics is static fatigue--i.e., a reduction in strength as a function of no other externally applied variable other than time (as opposed to mechanical or thermal fatigue). Quinn[43] reviews the published static fatigue data in generally early forms of silicon nitride and silicon carbide. It is pointed out that static fatigue is the general name for this observed time dependence, but that various origins and mechanisms can be involved, either acting singly or simultaneously operable. For instance, stress corrosion is environmentally assisted strength degradation in materials subjected to an externally applied stress. The mechanism involves the growth of cracks by chemical reaction with and attack by one or more elements or compounds present in the surrounding gaseous or liquid environment. The material at the existing crack tip is chemically changed and/or the local stress intensity is increased at the crack tip in an atomistic-level process. Alternatively, microcracks may nucleate (be created) in certain cases of stress corrosion.

Another form of static fatigue is creep rupture or creep fracture. This phenomenon refers to diffusion or cavitation-related deformation that results in the formation of extensive microcrack networks throughout the body. The microcracks eventually coalesce to form larger cracks which in time cause rupture at a lower value of stress than the fast fracture strength.

A third form of static fatigue is slow or subcritical crack growth (SCG). The concept of SCG involves the growth of pre-existing flaws under the applied stress, to the extent that the critical stress intensity at the crack tip is reached (thus leading to rapid fracture) at a lower macroscopic stress level than if the cracks were stable and unable to grow. The exact mechanism of SCG is related to material microstructure. For example, in many hot-pressed Si_3N_4 materials, the deformation of the intergranular phase leads to grain-boundary sliding. This promotes slow crack growth leading to lowered strength. The pre-exisiting flaws in HP-Si_3N_4 are often thought to be voids existing at intergranular triple points. Their extension is the mechanism to accommodate the grain-boundary sliding. As the flaws get larger, the strength, of course, decreases. Current materials research involves ways of achieving full densification with more deformation resistant intergranular phases.

Another form of strength degradation at elevated temperature, which can be termed static fatigue, is the generation of surface flaws by high temperature oxidation or gaseous corrosion mechanisms. This also includes the deposition of foreign elements on the material by the fuel. The concept here is that the intrinsic volume flaw population in the material is changed to a surface-related critical flaw population. This effect may or may not be related to the stress corrosion mechanism discussed above.

The following subsections of this report overview the elevated temperature strength observed for various silicon-base ceramics evaluated on this program. The observables are the strength and stress-strain behavior of the various materials. The existence or lack of strength or elastic modulus degradation is correlated with the nature of the fracture surfaces as viewed in the optical microscope, and the material impurities and microstructure. The results are usually interpreted assuming that slow crack growth is the predominant mechanism leading to strength reduction. This is readily observed on the fracture surfaces.

7.1 SILICON NITRIDE MATERIALS

The following sections overview the nature of high-temperature fracture in hot-pressed, sintered, and reaction-sintered forms of silicon nitride. Many of the basic differences in these materials were discussed in the final report of the predecessor to this program, AFML-TR-79-4188,[1] and presented at various conferences and symposia.[7,11,14,15,44]

7.1.1 Hot-Pressed Si_3N_4

The basis for the comparison of all hot-pressed Si_3N_4 materials is the behavior of Norton NC-132. This material is the most processing-mature, and was the most-utilized material in early component development programs. Its limitations form the basis for all of the strategies for improving high temperature properties through grain-boundary modification. During the time of the current program, significant developments have been accomplished. We can trace the need for such improvements and the path taken for realizing them by considering the behavior of the various HP-Si_3N_4 materials evaluated on this program.

(a) <u>MgO Additives</u>. The oxide additives used in hot-pressing Si_3N_4 to achieve full densification can result in the creation of intergranular phases in processed bodies that can deform readily at elevated temperature leading to strength reduction by subcritical crack growth. Evidence for subcritical crack growth in HP-Si_3N_4 is observed in both stress-strain data and directly on the fracture surfaces. Figure 23 illustrates the 4-point bend strength of HP-Si_3N_4 materials containing various amounts of MgO additives. The baseline here is the behavior of

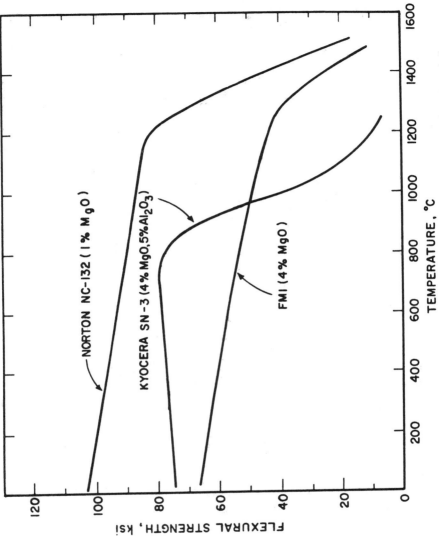

Figure 23. Flexural strength of hot-pressed silicon nitride materials.

Norton NC-132, which contains nominally 1% MgO. It is well documented in the ceramic literature how the presence of magnesia results in an amorphous intergranular phase in NC-132. Plasticity in the grain boundary silicate glass phase in NC-132 leads to strength reduction by SCG at temperatures T > 1250°C as shown in Figure 23. Evidence for increased plasticity at T > 1250°C is shown in the stress-strain behavior illustrated in Figure 24. The stress-strain curve is linear at 1200°C, and increasingly nonlinear with large increases in the strain-to-failure at 1350°C and 1500°C. The absence of SCG at 1200°C is confirmed by the appearance of the fracture surface, as shown in Figure 25. At higher temperatures SCG is observed on the fracture surfaces of NC-132, but the features of crack branching that indicate an operative slow crack growth mechanism are generally obscured by rapid oxidation at 1350°C and 1500°C. At 1500°C we have observed SCG to propagate through about 40% of the sample cross-section in NC-132.

With the behavior of NC-132, containing nominally 1% MgO, serving as a baseline, we can describe what happens when greater amounts of oxide additives are used to achieve densification. This applies, for instance, in the early development of a material, or in cases where the material is being developed for lower temperature applications, where economic considerations lead to larger amounts of oxide additives being used in processing. Kyocera, for instance, has developed a HP-Si_3N_4 with 4% MgO and 5% Al_2O_3 additives. Figure 23 illustrates that the low temperature strength is only maintained out to ~750°C. The stress-strain behavior correlates with this observation. Figure 26 illustrates that linear stress-strain behavior is exhibited for Kyocera SN-3 at 750°C, with distinctly nonlinear behavior obtained at 1000°C and above. Table 8 shows that this material contains ~10% oxide intergranular phase. Since this material is MgO-doped, it is expected that this phase is largely amorphous and deforms quite readily leading to strength reduction caused by SCG. Figure 27 shows the appearance of the fracture surface of

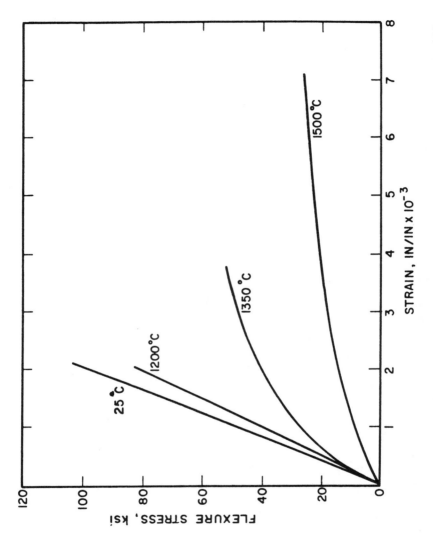

Figure 24. Representative flexural stress-strain behavior of Norton NC-132, HP-Si$_3$N$_4$ (1% MgO)

Materials Evaluation 107

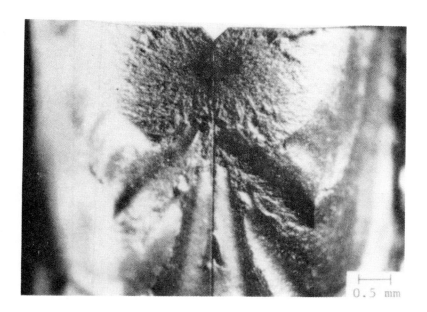

Figure 25. Fracture Surface (Tensile Surfaces Together) of Norton NC-132 HP-Si_3N_4 (1% MgO) Broken in Flexure at 1200°C.

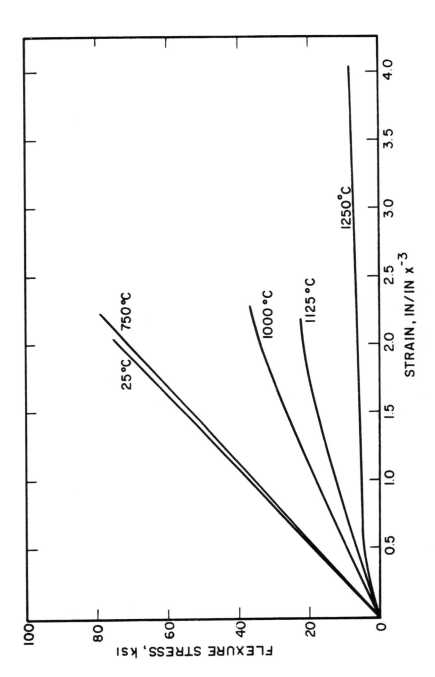

Figure 26. Flexural stress-strain behavior of Kyocera SN-3 HP-Si_3N_4 (4% MgO, 5% Al_2O_3).

Materials Evaluation 109

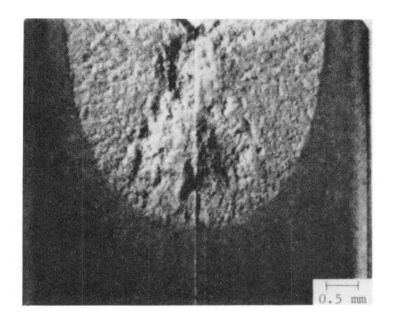

Figure 27. Fracture surface (tensile surfaces together) of Kyocera SN-3 HP-Si_3N_4 (4% MgO, 5% Al_2O_3) tested in flexure at 1125°C.

the Kyocera material tested at 1125°C. The extensive subcritical crack growth is evident. Similar reasoning can be used to explain the behavior of the FMI HP-Si_3N_4 containing nominally 4% MgO. The stress-strain behavior presented in Figure 28 illustrates distinctly nonlinear behavior at 1250°C.

(b) Y_2O_3 Additives. It was recognized by the early-to-mid 1970's that MgO additives resulted in amorphous magnesium silicate ("glassy") grain boundary phases in HP-Si_3N_4. We discussed the use of CeO_2 and Y_2O_3 as alternate additives in the final report on Contract F33615-75-C-5196 (AFML-TR-79-4188).[44] The graphical results are presented for early versions of such materials in Figures 29 and 30. The problem with these materials was generally that while good densification was achieved, a major problem existed at intermediate temperatures in the form of accelerated oxidation of one of the phases present, leading to material destruction. This is discussed in detail in Section 11, dealing with cumulative oxidation results to date.

A major emphasis in the Si_3N_4 technical community during the time of the present program was finding a suitable location in the Y_2O_3-SiO_2-Si_3N_4 phase diagram in which to process to avoid the oxidation instability problem. The Y_2O_3-modified HP-Si_3N_4 materials from Westinghouse and Toshiba evaluated on the present program illustrate the improvements that have been achieved.

Figure 31 illustrates that these two Y_2O_3-modified Si_3N_4 materials show much promise for improved high temperature strength. This is particularly evident for Toshiba HP-Si_3N_4 (4% Y_2O_3, 3% Al_2O_3). Figure 31 illustrates the improved strength at T > 1200°C for this material relative to NC-132. The stress-strain behavior for the Toshiba material shown in Figure 32 confirms the decreased intergranular plasticity when compared to NC-132. No evidence of SCG is visible on the fracture surface of Toshiba 4% Y_2O_3, 3% Al_2O_3-modified Si_3N_4 tested at 1350°C, as illustrated in Figure 33. Oxidation of this Y_2O_3, Al_2O_3-doped material was rapid enough at 1500°C to completely obscure the fracture

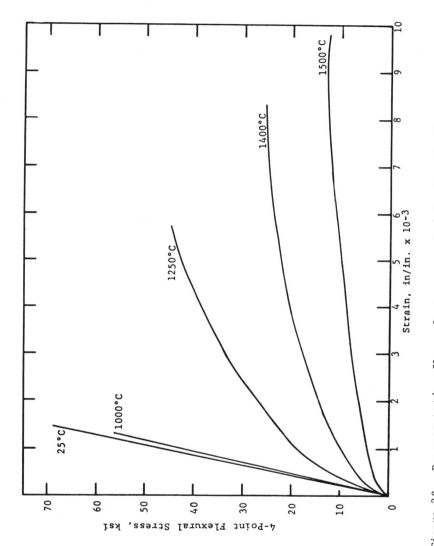

Figure 28. Representative flexural stress-strain behavior of Fiber Materials, Inc. HP-Si_3N_4 (4% MgO)

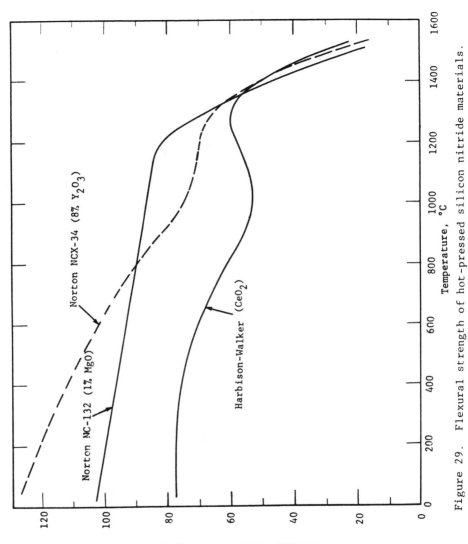

Figure 29. Flexural strength of hot-pressed silicon nitride materials.

Materials Evaluation 113

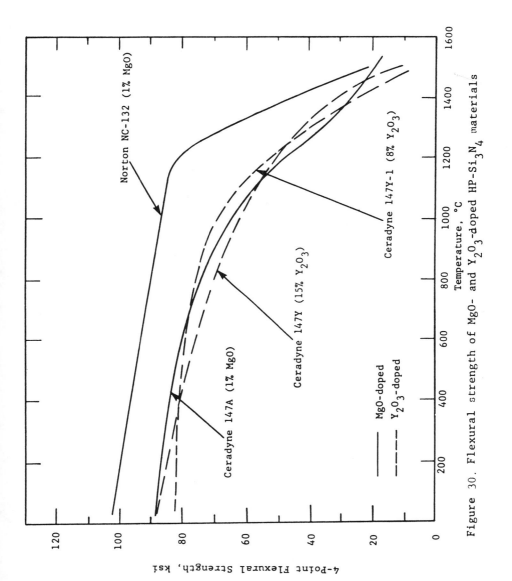

Figure 30. Flexural strength of MgO- and Y_2O_3-doped HP-Si_3N_4 materials

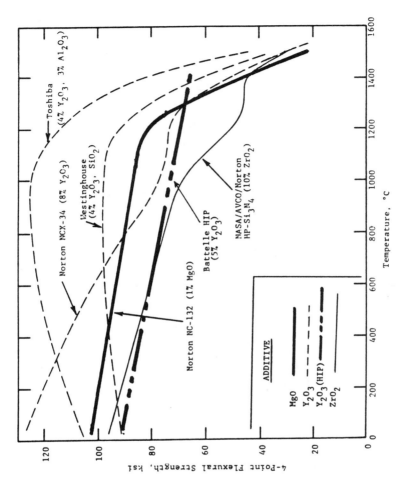

Figure 31. Flexure strength of various HP- and HIP-MgO, Y_2O_3, and ZrO_2-containing silicon nitride materials.

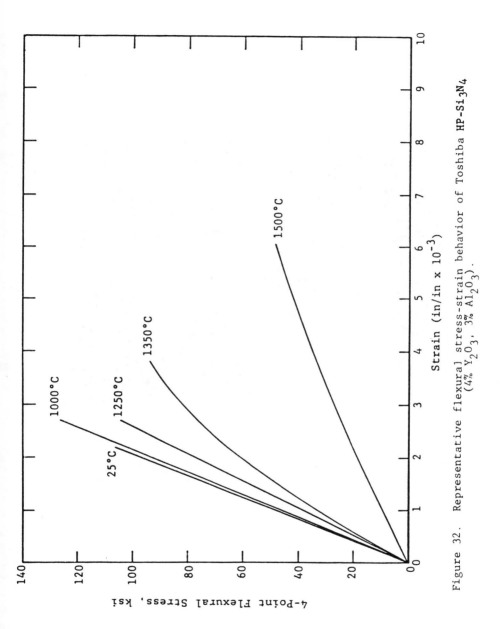

Figure 32. Representative flexural stress-strain behavior of Toshiba HP-Si$_3$N$_4$ (4% Y$_2$O$_3$, 3% Al$_2$O$_3$).

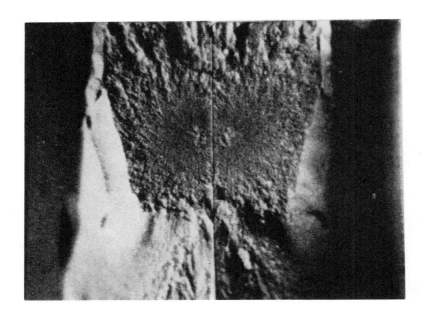

Figure 33. Fracture surface (tensile surfaces together) of Toshiba HP-Si_3N_4 (4% Y_2O_3, 3% Al_2O_3) broken in flexure at 1350°C (sample T01F30).

features. However, the oxidation rate of Y_2O_3, SiO_2-modified Si_3N_4 structures is lower, and the fracture features at both 1350° and 1500°C are readily observed for Westinghouse HP-Si_3N_4, as shown in Figure 34. This material exhibited improved strength when compared to NC-132 (Figure 31), with no detectable SCG at 1350°C. At 1500°C, some SCG is visible on the fracture surface, as shown in Figure 34. These observations correlate with the shape of the stress-strain curve for the Westinghouse material, presented in Figure 35. Note that at 1350°C nearly linear behavior was obtained, whereas the stress-strain relationship is distinctly non-linear at 1500°C.

The improved elevated temperature strength of the Westinghouse and Toshiba materials is the result of the use of Y_2O_3 as a densification additive. Y_2O_3 dopants result in an oxide intergranular phase that can be crystallized by post-densification heat treatment. This results in more deformation resistant grain boundaries, and less grain boundary sliding, resulting in less subcritical crack growth. The stress-strain behavior and appearance of the fracture surfaces confirm this. X-ray diffraction studies, however, were inconclusive in attempts to identify crystallinity within intergranular regions. Higher resolution techniques such as transmission electron microscopy are required to detect the presence of a crystalline intergranular phase.

These results for the Westinghouse and Toshiba materials using Y_2O_3 as a processing additive illustrate the potential of ceramics to be successfully utilized in structural high temperature applications. It is judged that the Westinghouse material has extended the temperature limit approximately 100°C beyond that of MgO-doped NC-132 HP-Si_3N_4. The ~90 ksi room-temperature strength is maintained out to 1250°C, where linear stress-strain behavior is still observed. At 1350°C, the strength drops to ~63 ksi, but there is only a slight departure from stress-strain linearity, and SCG is not visible optically on fracture surfaces. The Toshiba HP-Si_3N_4, modified by 4% Y_2O_3 and 3% Al_2O_3 addition, also extends the potential use temperature of silicon nitride out

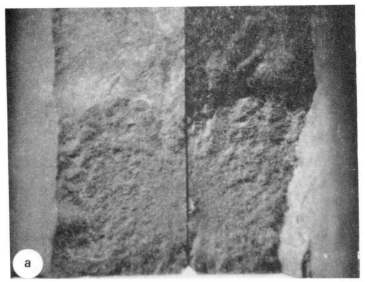

16X

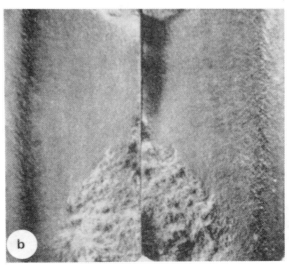

12X

Figure 34. Fracture surfaces (tensile surfaces together) of Westinghouse HP-Si_3N_4 (4% Y_2O_3, SiO_2) tested at 1350° and 1500°C.

(a) 1350°C

(b) 1500°C

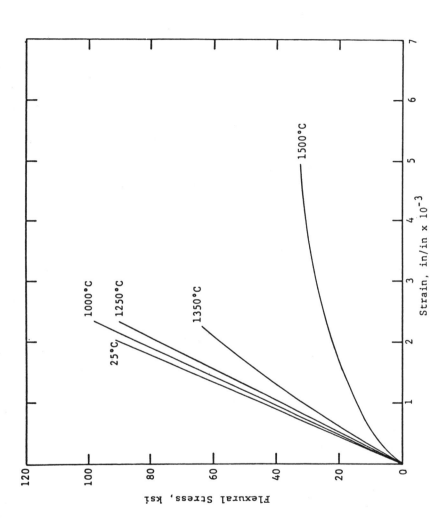

Figure 35. Representative flexural stress-strain behavior of Westinghouse HP-Si_3N_4.

to 1350°C. This material exhibited a bend strength of 94 ksi at this temperature, with nearly linear stress-strain, and no optically discernible evidence of subcritical crack growth on the fracture surfaces. However, the Toshiba HP-Si_3N_4 (4% Y_2O_3, 3% Al_2O_3) appears to be oxidation-limited at 1500°C. This is thought to be due to the Al_2O_3 additive.

These results are encouraging. They demonstrate that Y_2O_3 additions can be used to achieve a deformation-resistant intergranular phase that results in improved properties at elevated temperature. The Toshiba material achieved this through the use of Y_2O_3 and Al_2O_3 additives. The Westinghouse material achieved this through the use of Y_2O_3 and SiO_2 additives. Other oxide additives, or combinations of successful ones, however, can have an adverse effect on high temperature strength. For instance, Figure 36 illustrates another Toshiba material, which was developed for a lower temperature application. It exhibited rapid strength degradation after T ~1000°-1100°C. This material contained 3% Y_2O_3, 4% Al_2O_3, and an undetermined amount of SiO_2. The stress-strain behavior shown in Figure 37 further illustrates that degradation occurs at above 1100°-1200°C. Note the extremely nonlinear behavior at 1350°C. Figure 38 confirms that significant crack-branching was observed on the fracture surfaces at temperatures as low as 1250°C. Table 8 showed that almost twice as much oxygen was present compared to the other Toshiba material. Apparently SiO_2 was present in significant quantities, and either resulted in amorphous silicate grain boundary phases or inhibited the crystallization of yttrium silicate intergranular phases. However, the presence of SiO_2 does increase the oxidation resistance of Y_2O_3 and Y_2O_3-Al_2O_3 modified silicon nitride materials.

The success of Y_2O_3 as a densification aid for HP-Si_3N_4 lies in the fact that the resulting yttrium silicate intergranular phase can be crystallized. If more than 4% Y_2O_3 is used (i.e., 8% or more), we have found that there is a strong tendency to be in that part of the Si_3N_4-Y_2O_3-SiO_2 phase triangle that results

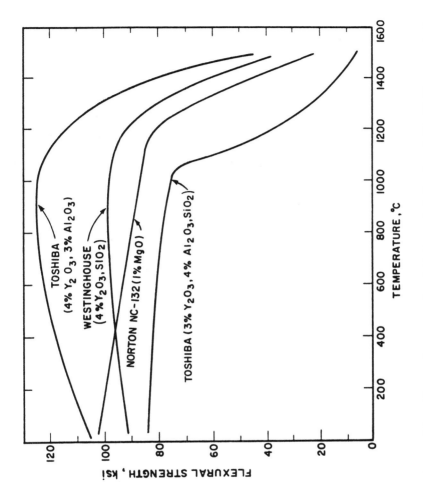

Figure 36. Flexural strength of MgO- and Y_2O_3-doped HP-Si_3N_4 materials.

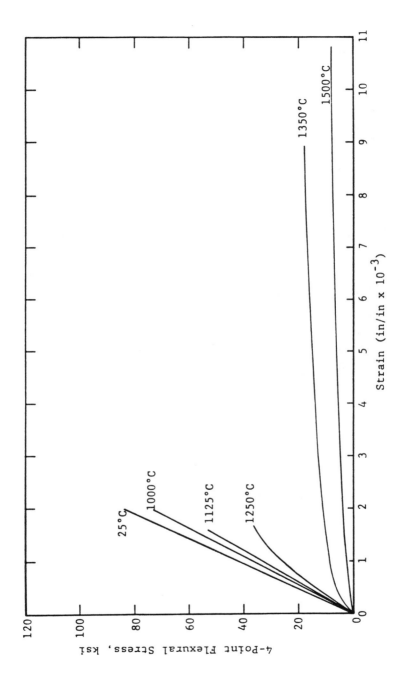

Figure 37. Representative flexural stress-strain behavior of Toshiba HP-Si_3N_4 (3% Y_2O_3, 4% Al_2O_3, SiO_2).

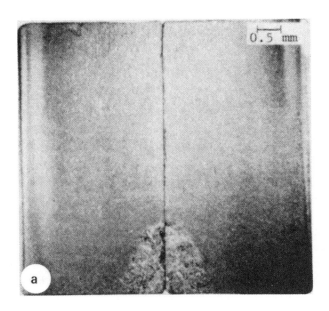

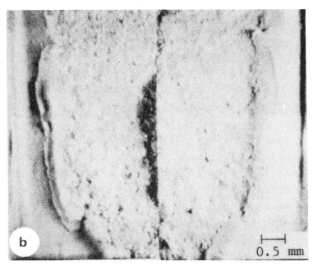

Figure 38. Fracture surface (tensile surfaces together) of Toshiba HP-Si_3N_4 (3% Y_2O_3, 4% Al_2O_3, SiO_2) broken in flexure at 1250°C and 1350°C.

(a) 1250°C

(b) 1350°C

in oxynitride phases that are unstable in oxidizing environments. These phases experience accelerated oxidation, with resulting macrocracking of the ceramic body. This will be discussed in detail in Section 11, Oxidation.

(c) <u>HIP-Si_3N_4 (Y_2O_3)</u>. The use of Y_2O_3 densification additives for Si_3N_4 discussed above illustrates that intergranular crystallinity greatly impedes the relative movement of grains, thereby reducing the phenomenon of slow crack growth accommodated by cavity nucleation, extension of pre-existing triple-point voids, etc. There is one other route to decreasing SCG being investigated by several laboratories: hot isostatic pressing. HIP'ing is performed at much higher pressures than can be achieved in uniaxial hot pressing (e.g., 30,000 psi). Thus, the Si_3N_4 powder can be compacted to full density using a lower concentration of oxide additive, thereby minimizing slow crack growth, and improving the elevated temperature strength and creep resistance. HIP'ing has the additional economic advantage of the potential for near-net-shape fabrication of complex components.

A small quantity of Battelle HIP-Si_3N_4 containing nominally 5% Y_2O_3 was available for evaluation on this program. This material was prepared by isopressing Si_3N_4 powder at 30,000 psi, followed by HIP'ing at 1725°C for 1 hr at 30,000 psi.[45] The material was reported to consist of submicron equiaxed Si_3N_4 grains and a thin amorphous intergranular film that is very refractory. Battelle compressive creep data for this material are compared to flexural creep data for various Y_2O_3-containing HP-Si_3N_4 evaluated on the present program in Figure 39. This material exhibits low creep rates and shows much promise. Accordingly, time-dependent flexure tests were conducted on the present program at 1400°C in air. In this test the fracture stress is compared with the time-to-failure. These tests are sometimes referred to as dynamic fatigue or differential strain rate tests, and yield a quantitative measure of the time-dependent strength degradation by the phenomenon of subcritical crack growth. These tests are essentially regular flexure tests where the

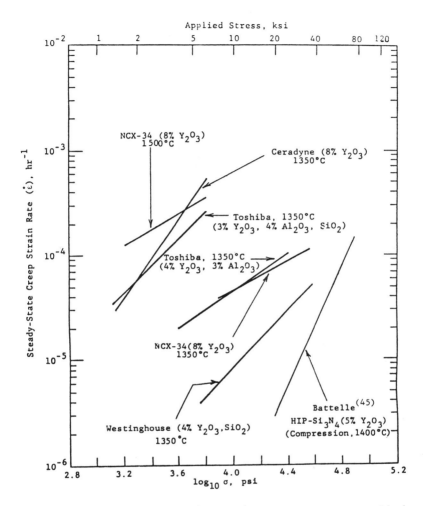

Figure 39. Steady-state flexural creep rate vs. applied stress for various Y_2O_3-modified Si_3N_4 materials.

time-to-failure is varied by adjusting the testing machine crosshead speed. Regression analysis of the resulting fracture strength vs. time relation yields a value for n in the relation v = AK^n, where v is the crack velocity, A is a constant, and K is the stress intensity at the crack tip. A large value of n indicates very little subcritical crack growth, and thus time-invariant strength.

The results of these tests are plotted in Figure 40 along with (a) Battelle data on the same material, tested at 1400°C, but at a machine crosshead speed a factor of 25 faster (0.005 ipm), and (b) our fast fracture results (crosshead speed 0.02 ipm) for various other Y_2O_3-containing HP-Si_3N_4 materials evaluated on this program. It is seen that the Battelle HIP-Si_3N_4 (5% Y_2O_3) material has a room-temperature strength of ~90 ksi. The 1400°C strength of the HIP'ed material is very good, ~63 ksi. The present IITRI data agree with the Battelle generated strength. This material has a 1400°C fracture strength about the same as we previously found for Westinghouse HP-Si_3N_4, which is doped with nominally 4% Y_2O_3 and an undetermined amount of SiO_2. What is significant here is that the Battelle and IITRI strengths were measured at crosshead deformation rates a factor of 25 and 100, respectively, slower than the Westinghouse and other materials, as shown in Figure 40. This means that there is apparently very little subcritical crack growth occurring prior to fracture in the HIP'ed material (however, the Westinghouse material was not subjected to the slow (0.0002 ipm) flexure test, so no direct comparison can be made).

Figure 41 illustrates the fracture surfaces of the Battelle HIP-Si_3N_4 (5% Y_2O_3) material after slow 1400°C flexure testing, where the time-to-failure was typically 113 min. We see only slight evidence of crack branching on the fracture surfaces, the SCG region extending over about 3% of the test bar cross-section. Battelle[45] indicated that it was reasonable to assume that this material would exhibit some SCG prior to fracture at 1400°C, but they could not detect any in optical examination of the fracture

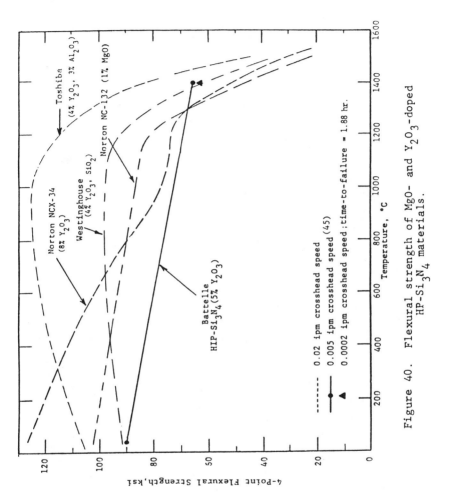

Figure 40. Flexural strength of MgO- and Y_2O_3-doped HIP-Si_3N_4 materials.

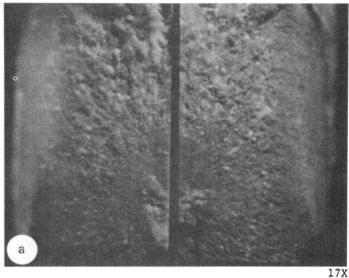

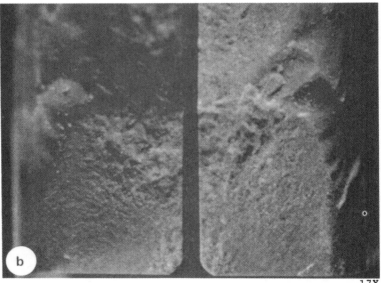

Figure 41. Fracture surfaces of Battelle HIP-Si$_3$N$_4$ (5% Y$_2$O$_3$) after time-dependent flexure test at 1400°C. (The time-to-failure was typically 113 min; the machine crosshead speed was 0.0002 ipm.)

(a) Sample BW1F1
(b) Sample BW1F2

surfaces. However, their test was conducted at a deformation rate a factor of 25 faster than the present IITRI tests. The time-to-failure in the Battelle test was ~5 min (by IITRI estimate, as no time-to-failure data were reported by Wills and Brockway). Thus, the IITRI and Battelle results are consistent: very good strength retention at 1400°C, and only slight evidence of SCG phenomena when the fracture times approach 2 hr.

It is interesting to compare our present slow-flexure results on this HIP'ed material with results we have obtained for Norton NC-132 HP-Si_3N_4 (1% MgO) and Norton NCX-34 HP-Si_3N_4 (8% Y_2O_3). In this case, direct comparison of the materials is possible since all were flexure-tested at a crosshead deformation rate of 0.0002 ipm, which resulted in fracture times on the order of 2-3 hr. The comparison of these materials is presented in Figure 42, where fracture stress is plotted against time-to-failure. It is seen that Norton MgO- and Y_2O_3-doped HP-Si_3N_4 materials experienced significant strength degradation at 1400°C, the degree of subcritical crack growth being characterized by n values of 16 and 12, respectively. In contrast, the 1400°C n value of Battelle HIP-Si_3N_4 (5% Y_2O_3) was estimated to be much higher, n ~30, which indicates much less SCG.

The fracture surface features of these three materials are consistent with this interpretation. The appearance of the fracture surface of NC-132 HP-Si_3N_4 (1% MgO) after 0.0002 ipm flexure testing is presented in Figure 43. The strength of NC-132 was 33% lower than the 1400°C fast fracture strength, the time-to-failure was typically 44 min, and slow crack growth extended over ~20% of the sample cross-section. For NCX-34 HP-Si_3N_4 (8% Y_2O_3), the strength for the same test conditions was 47% lower than the 1400°C fast fracture strength; but the failure time was typically 206 min, considerably longer than for NC-132. Figure 44 illustrates that the degree of SCG was lower in NCX-34, the SCG markings only extending over ~10% of the bend bar transverse section. These two fracture surfaces for NC-132 and NCX-34 are compared to the fracture surface for Battelle HIP-Si_3N_4 (5% Y_2O_3) that was

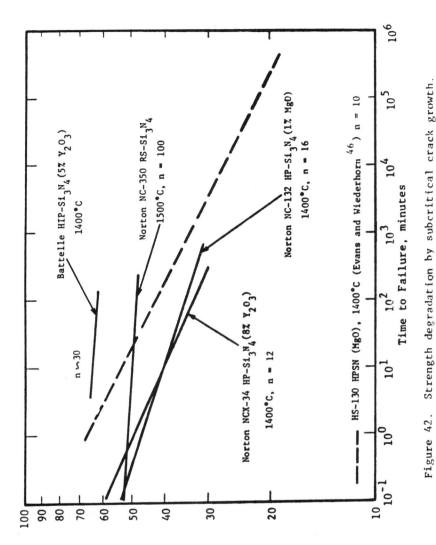

Figure 42. Strength degradation by subcritical crack growth.

Materials Evaluation 131

12X

Figure 43. Fracture surface of Norton NC-132 HP-Si$_3$N$_4$ (1% MgO) after time-dependent flexure test at 1400°C. (The strength was 33% lower than the 1400°C fast fracture strength; the time-to-failure was typically 44 min; the machine crosshead speed was 0.0002 ipm.)

12X

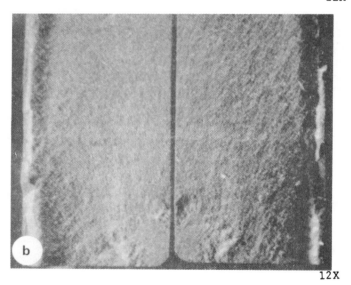

12X

Figure 44. Fracture surfaces of Norton NCX-34 HP-Si$_3$N$_4$ (8% Y$_2$O$_3$) after time-dependent flexure test at 1400°C. (The strength was 47% lower than the 1400°C fast fracture strength; the time-to-failure was typically 206 min; the machine crosshead speed was 0.0002 ipm.)

(a) Sample 51F60
(b) Sample 51F58

presented in Figure 41. As discussed above, SCG features only extended over ~3% of the cross-sectional area of the HIP'ed material.

Therefore, the present results indicate that the degree of subcritical crack growth and associated strength degradation can be substantially reduced by utilizing the high pressure attainable with the HIP'ing process to attain full density with a lower concentration of oxide additive. The ultimate behavior goal might be to achieve the time-invariant strength of additive-free reaction-sintered Si_3N_4.* Figure 42 illustrates the results for Norton NC-350 RS-Si_3N_4 in the slow flexure tests. The test temperature was 1500°C (100°C higher than for the HP- and HIP-Si_3N_4 materials), and for all practical purposes the strength did not vary with time (n = 100). This goal may not be reached for hot-pressed or HIP'ed Si_3N_4, because some small amount of oxide additive may always be necessary to achieve full density in pressure-assisted sintering of Si_3N_4 powder.

(d) $\underline{ZrO_2\text{-containing } Si_3N_4}$. Yttria and ceria are not the only potential additives for HP-Si_3N_4 being investigated. It has been thought for some time that zirconia compounds in intergranular regions would be quite resistant to deformation. Furthermore, zirconia-based materials might not exhibit the intermediate-temperature phase instabilities often associated with yttria additives.

NASA-Lewis supplied a HP-Si_3N_4 (10% ZrO_2)** material to this program that has proven to be quite interesting. The material

---

*This comment is made only with respect to the absence of inter-granular SCG at high temperature. Certainly, it is recognized that the denser forms of Si_3N_4 have advantages over porous RS-Si_3N_4 such as higher low-temperature strength and lower surface area. Internal oxidation is one of the major disadvantages of reaction-sintered Si_3N_4.

**AFWAL-developed Zyttrite, a fully stabilized ZrO_2 containing 6 mol% (10.8 wt%) Y_2O_3.

concept was developed by Vasilos et al[47] at AVCO in an attempt to improve the stress rupture behavior and creep strength of Si_3N_4 by using ZrO_2 additives to create a zirconium oxynitride grain boundary phase. The actual material evaluated in this program was fabricated at Norton Company, using the same starting powder and procedures used for NC-132. We have thus termed this material NASA/AVCO/Norton HP-Si_3N_4 (10% ZrO_2). Our verification of its improved creep properties will be presented in Section 8.

Figure 31 illustrated that the high-temperature strength of this material is generally lower than that of the other high-performance Si_3N_4 materials shown. However, fracture origins at all elevated temperatures were the same shiny dark inclusion particles that were the sources of fracture at room temperature. Thus, the NASA/AVCO/Norton material is distinctly different from the others. Its elevated-temperature strength is controlled by the same inclusion-related fracture origins that were dominant at 25°C. There is no evidence of subcritical crack growth in this material, even at 1500°C in the fast fracture test. Evidence for this is shown in Figure 45b. The optical photograph of the fracture surface reveals no observable markings that would indicate operative slow crack growth (SCG). Note also in Figure 45b that the 1500°C fracture surface features are not obscured at all by oxidation. Thus, it can be expected that ZrO_2 additives lead to superior oxidation resistance when compared to other oxide additives (such as MgO). For comparison, the 1500°C fracture surface of the Westinghouse HP-Si_3N_4 (4% Y_2O_3, SiO_2) material was shown in Figure 34b. Note the region of SCG extending across ~20% of the bend bar cross section. Figure 31 illustrates that the Toshiba HP-Si_3N_4 (4% Y_2O_3, 3% Al_2O_3) was the strongest Si_3N_4 evaluated on this program. Its fracture surfaces were badly obscured by oxidation at 1500°C. Thus, no estimate of the extent of SCG was possible (no attempts were made to remove the oxide, such as by soaking in hydrofluoric acid).

The stress-strain behavior for the NASA/AVCO/Norton material also indicates the presence of a deformation-resistant grain

Materials Evaluation 135

24X

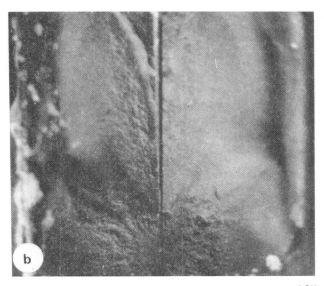

12X

Figure 45. Fracture surfaces of NASA/AVCO/
Norton HP-Si$_3$N$_4$ (10% ZrO$_2$).

(a) Sample NAlF5 tested at room temperature
(b) Sample NAlF14 tested at 1500°C

boundary phase. Figure 46 illustrates linear behavior in the limited testing at 1500°C. This is the first evidence of such behavior that has been observed for any hot-pressed Si_3N_4 evaluated on this program. The stress-strain behavior of the Y_2O_3-doped Toshiba and Westinghouse materials was presented in Figures 32 and 35, respectively. Note the nonlinearity at 1500°C, especially for the Westinghouse material.

The linear stress-strain behavior and the absence of observable SCG markings on the fracture surfaces of the NASA-AVCO/Norton HP-Si_3N_4 (10% ZrO_2) material in 1500°C fast fracture tests illustrate that this material has a very refractory intergranular phase. This will explain the excellent creep resistance for this material which is reported in Section 8. It would be expected, then, that this material would exhibit noticeably better time-dependent flexure strength and stress rupture behavior as well. Accordingly, fast and slow flexure strength tests were conducted at 1400°C, and the time-to-failure was recorded. The results of these constant stress rate tests are summarized in Figure 47, which illustrates the fracture stress vs. time for various silicon nitride materials evaluated on this program. The slope of the curves is -1/n, where n is the exponent in the $v = AK^n$ relation describing crack velocity vs. stress intensity. A high value of the n parameter in Figure 47 indicates less time dependence of the strength, and thus less subcritical crack growth. It is observed that the NASA/AVCO/Norton material exhibits the highest n value of any HP-Si_3N_4 material evaluated on this program (n = 36). This agrees with the fast fracture results discussed above. This material exhibits surprisingly little SCG and resultant strength degradation at elevated temperature. The fracture surface of a typical NASA/AVCO/Norton HP-Si_3N_4 (10% ZrO_2) material broken in slow flexure is shown in Figure 48. Note the absence of SCG markings. Corresponding fracture surfaces of the other hot-pressed silicon nitride materials compared in the stress-time plot of Figure 47 were shown in Figures 41, 43, and 44. The amount of SCG seen on the fracture surfaces

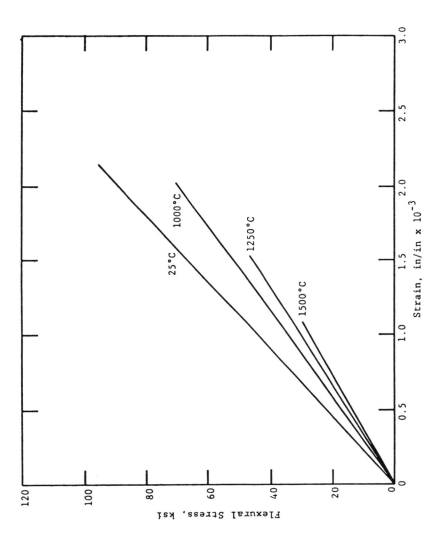

Figure 46. Flexural stress-strain behavior of NASA/AVCO/Norton HP-Si_3N_4 (10% ZrO_2).

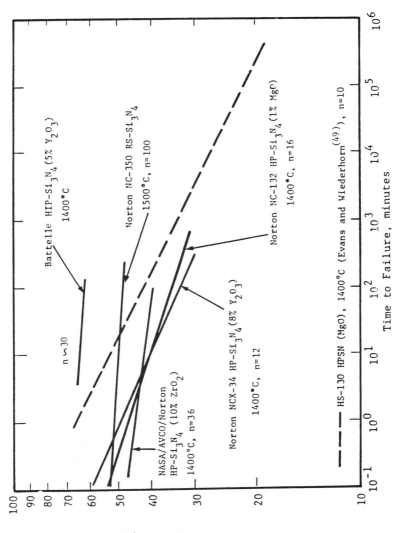

Figure 47. Strength degradation by subcritical crack growth.

12X

Sample NA1F40

Figure 48. Fracture surface of NASA/AVCO/Norton HP-Si$_3$N$_4$ (10% ZrO$_2$) after time-dependent flexure test at 1400°C. (The strength was typically 13% lower than the 1400°C fast fracture strength; the time-to-failure was typically 90 min; the machine crosshead speed was 0.0002 ipm.)

correlates with the amount of strength degradation with time seen in Figure 47.

Since the NASA/AVCO/Norton material exhibited no noticeable SCG in fast fracture tests, and very good behavior in the slow flexure tests where the failure times were typically 1-2 hr, it was of interest to evaluate this material at much longer times. Accordingly, flexural stress rupture tests were conducted, where a constant stress was applied, and the time to failure was recorded. The tests were conducted at 1400°C in air atmosphere. The results are summarized in Figure 49. It is observed that at an applied stress of 30 ksi (65% of the 1400°C fast fracture strength) sample failures were not obtained in times as long as over 1700 hr. Even when fracture was obtained, for instance at 209 hr at an applied stress of 35 ksi, Figure 50 illustrates the absence of significant SCG. This is a significant achievement in hot-pressed silicon nitride technology, and indicates the advantage of using ZrO_2 as an additive. In contrast, Figure 47 illustrates the much more rapid strength degradation of MgO-doped HP-Si_3N_4.

In summary, it has been shown that the NASA/AVCO/Norton HP-Si_3N_4 (10% ZrO_2) material exhibits superior high-temperature properties. This is the result of the zirconium oxynitride intergranular phase that results in significantly reduced slow crack growth and improved time-dependent strength.

7.1.2 Reaction-Sintered Si_3N_4

At the other end of the silicon nitride behavioral spectrum is reaction sintered (RS) Si_3N_4. Figures 51 and 52 illustrate the strength-temperature behavior of typical high performance RS-Si_3N_4 materials. The room-temperature strength is only 30-40 ksi at best, caused by the presence of 10-20% porosity. Fracture origins are typically large pores and pore agglomerates (Figure 53). However, RS-Si_3N_4 materials are very pure, and thus their strength is maintained at elevated temperatures (any strength increase observed at high temperature is thought to be caused by

Materials Evaluation   141

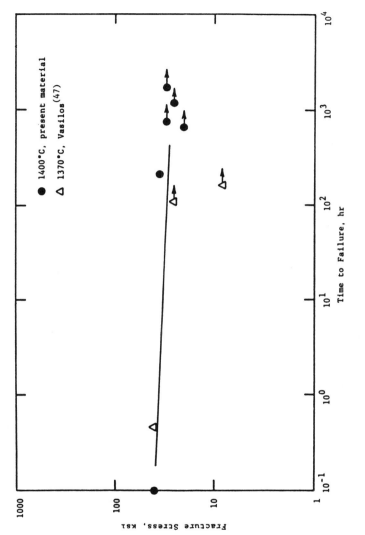

Figure 49. Flexural stress-rupture of NASA/AVCO/Norton HP-Si$_3$N$_4$ (10% ZrO$_2$) at 1400°C (air atmosphere).

Figure 50. Fracture surfaces of NASA/AVCO/ Norton HP-Si_3N_4 (10% ZrO_2) after 1400°C stress rupture test in air atmosphere (the applied stress was 35 ksi; the sample failed at 209 hr; the sample was soaked in HF to remove the oxide scale to better observe the fracture surface markings).

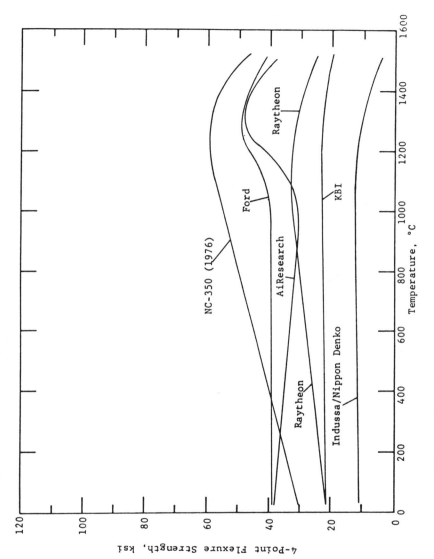

Figure 51. Flexure strength of reaction sintered Si_3N_4.

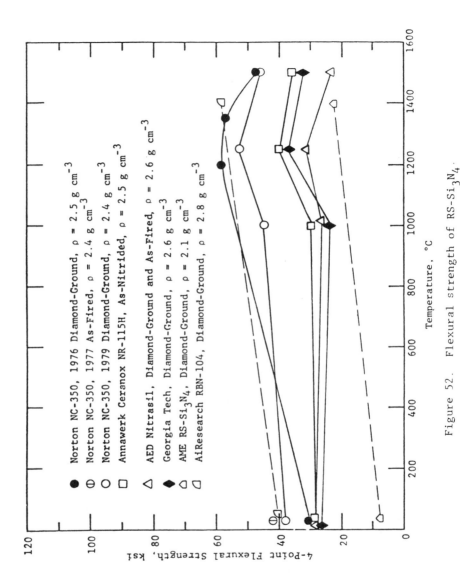

Figure 52. Flexural strength of RS-Si_3N_4.

Materials Evaluation 145

12X

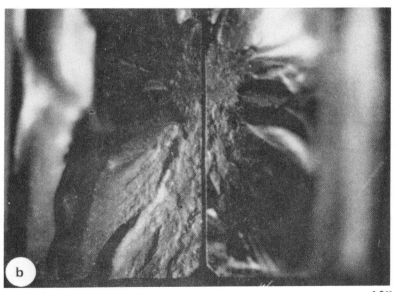

12X

Figure 53. Fracture surfaces (tensile surfaces together) of AiResearch RBN-104 RS-Si_3N_4 tested at room temperature and 1400°C.
(a) Sample tested at room-temperature
(b) Sample tested at 1400°C

oxidation, i.e., blunting and rounding of sharp pore features). Indeed, Table 8 shows that the oxygen content of RS-Si_3N_4 materials is very low. Typical stress-strain curves are given in Figure 54. Note that linear behavior is obtained at 1500°C. Figure 55 shows the fracture surface appearance of Norton NC-350 RS-Si_3N_4 tested at 1500°C in air atmosphere. Note the absence of SCG markings on the fracture surface. Therefore, the absence of oxide intergranular phases is the reason for the excellent elevated temperature strength retention in RS-Si_3N_4.

7.1.3 Sintered Si_3N_4

Sintered Si_3N_4 is similar to hot-pressed silicon nitride in that oxide additives are employed to achieve consolidation. Analagous to HP-Si_3N_4, this is accomplished somewhat at the expense of elevated temperature properties. However, sintered Si_3N_4 has one distinct advantage--the capability to produce net shapes of rather complex configuration. Only a few SSN materials were evaluated on this program. The results are summarized in Figures 56 and 57.

7.2 SiC MATERIALS

The effect of intergranular phases resulting from processing additives on subcritical crack growth and strength reduction is much less pronounced in SiC than it is in Si_3N_4. Figure 58 illustrates the strength-temperature behavior of the hot-pressed SiC materials evaluated on this program. The most successful HP-SiC contains ~2% Al_2O_3 as a densification aid. This results in aluminosilicate intergranular phases, the deformation of which is partially responsible for the strength reduction in HP-SiC at elevated temperature. Boron carbide, B_4C, results in a more deformation-resistant grain boundary. This is seen in Figure 58. However, as discussed in Section 6.2 above, B_4C promotes growth of SiC grains, which results in comparatively low strength.

With the successful development of fully dense sintered SiC, comparatively little work continues on HP-SiC. Sintered SiC

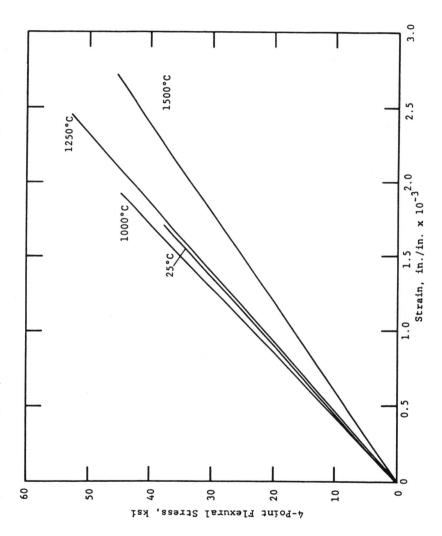

Figure 54. Representative flexural stress-strain behavior of Norton NC-350 RS-Si_3N_4 (1979).

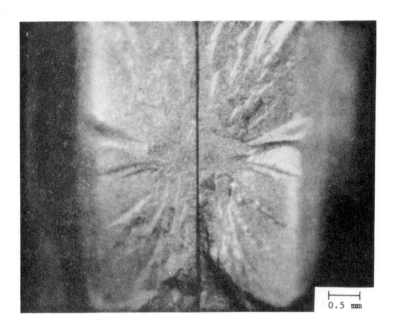

Figure 55. Fracture surface (tensile surfaces together) of Norton NC-350 RS-Si_3N_4 tested in flexure at 1500°C.

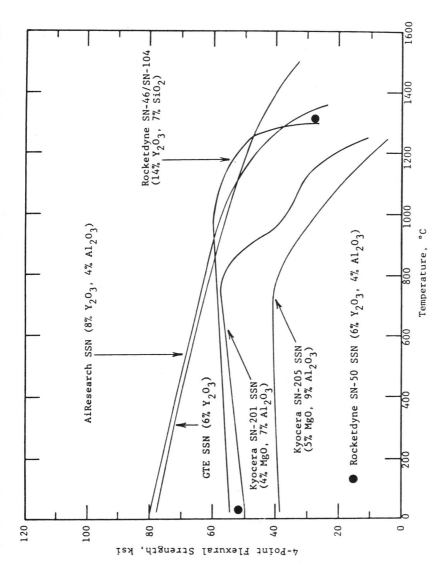

Figure 56. Flexural strength of sintered silicon nitride materials.

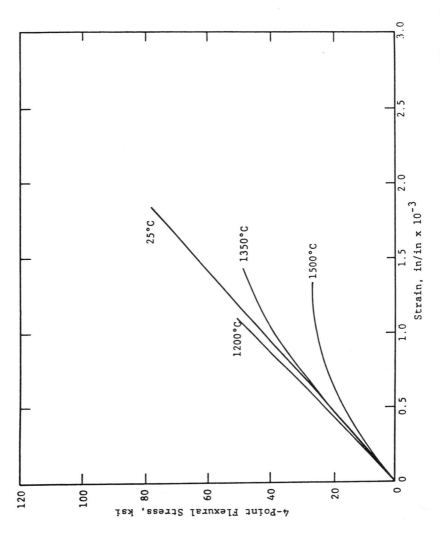

Figure 57. Representative flexural stress-strain of GTE sintered Si_3N_4 (6% Y_2O_3).

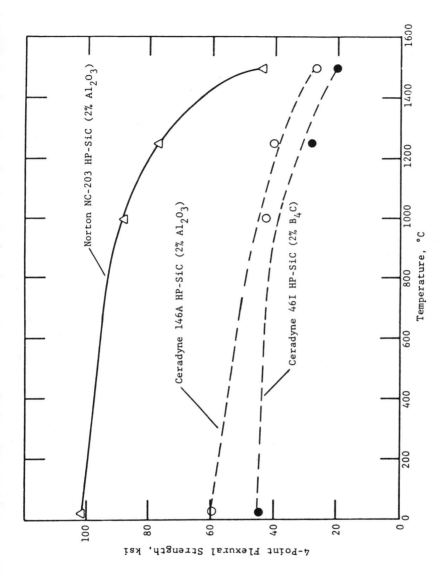

Figure 58. Flexural strength of hot-pressed silicon carbide materials.

offers the advantage of the fabricability of complex shapes. This technology has advanced by the incorporation of carbon and boron that permit sintering to full density. Furthermore, these additives do not result in elevated temperature strength degradation like the additives for silicon nitride. Figure 59 illustrates the strength-temperature behavior of sintered SiC, which contains few additives, if any. This is in contrast to that of hot-pressed SiC, which contains 1-2% Al_2O_3 additive to achieve densification, but which results in aluminosilicate grain boundary phases. The low temperature strength is maintained much better out to 1500°C for the sintered materials. Sintered SiC contains very little residual oxide impurity. Figures 60 and 61 (for the Carborundum Hexoloy SX-05 and Kyocera SC-201) show, analagous to high purity RS-Si_3N_4, that linear stress-strain behavior is obtained in sintered SiC at 1500°C. Figure 62 illustrates typical fracture surfaces for sintered SiC tested at 1000°, 1250°, and 1500°C. No fracture surface features are visible (at this level of magnification) that would indicate an operative slow crack growth mechanism. Oxide additives in HP-SiC do not affect the strength as much as they do in hot-pressed Si_3N_4. Figure 63 illustrates that only slightly nonlinear behavior is observed in 1500°C stress-strain data for Norton NC-203 HP-SiC, which contains ~2% Al_2O_3. Most HP-Si_3N_4 materials exhibit much more pronounced nonlinear stress-strain behavior at 1500°C. For the same reason, the temperature at which the strength begins to decrease precipitously is lower for Si_3N_4 (~1250°C). The reason for the less pronounced temperature dependence observed for SiC may be due to a combination of effects: (a) the absence of significant intergranular phases in sintered SiC, (b) the aluminosilicate intergranular phase in HP-SiC being relatively refractory, and (c) the fracture mode in SiC being predominantly transgranular, rather than intergranular as in Si_3N_4. A summary of the elevated temperature elastic properties of various sintered SiC materials evaluated on this program is presented in Table 17.

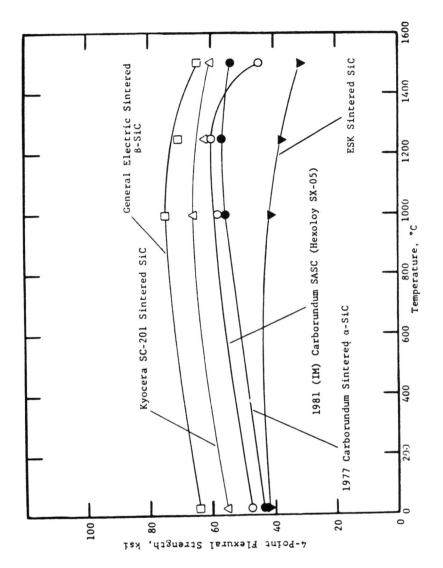

Figure 59. Flexural strength of sintered silicon carbide materials.

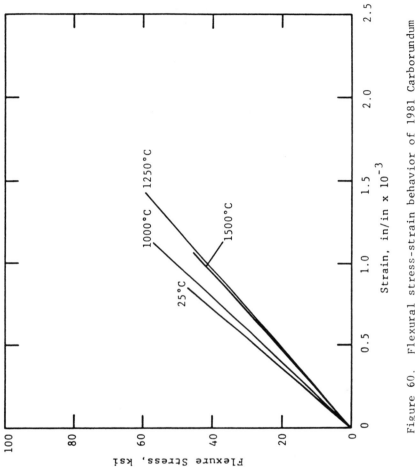

Figure 60. Flexural stress-strain behavior of 1981 Carborundum sintered α-SiC (Hexoloy SX-05).

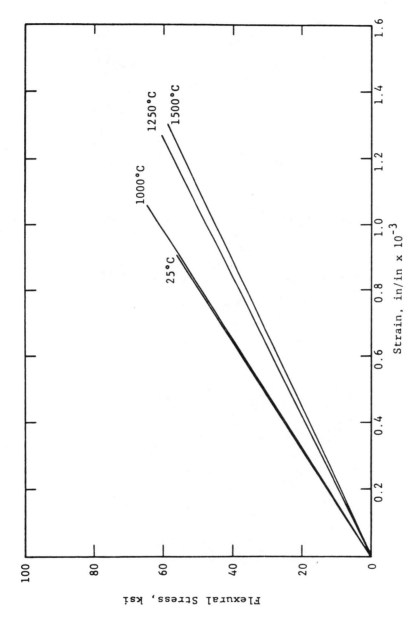

Figure 61. Representative flexural stress-strain behavior of Kyocera SC-201 sintered SiC.

156 Ceramic Materials for Advanced Heat Engines

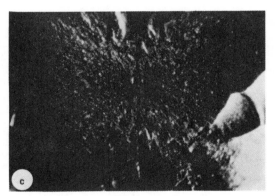

Figure 62. Fracture surfaces (tensile surfaces together) of Kyocera SC-201 sintered SiC and 1977 Carborundum sintered α-SiC.

(a) Kyocera SC-201 sintered SiC tested at 1000°C.
(b) Kyocera SC-201 sintered SiC tested at 1250°C.
(c) 1977 Carborundum sintered α-SiC tested at 1500°C.

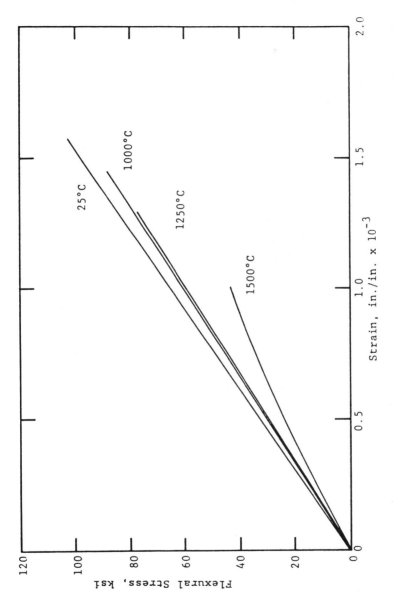

Figure 63. Representative stress-strain behavior for Norton NC-203 HP-SiC (2% Al_2O_3).

Table 17
ELASTIC MODULUS OF SINTERED SiC

Temp., °C	Elastic Modulus, 10^6 psi[a]			
	Carborundum 1977 Slip Cast (α)	Carborundum 1981 Injection Molded (α)	General Electric Boron-Doped (β)	Kyocera SC-201 (α)
25	58.2	55.9	54.8	60.5
1000	67.0	51.1	63.2	60.5
1250	63.3	40.8	57.1	50.1
1500	54.7	42.1	58.3	44.9

[a] Secant Young's modulus determined using electromechanical deflectometer to continuously read outer fiber tensile strain during a 4-point bend test at a machine crosshead speed of 0.02 ipm in air atmosphere. The values tabulated are as-measured, and not corrected for porosity.

TABLE 18. ELASTIC MODULUS OF COORS SILICONIZED SiC

Temperature, °C	Elastic Modulus, 10^6 psi[a]		
	1979 SC-1	1981 SC-2	1982 SC-2
25	51.1	53.8	54.2
1000	42.6	55.5	55.5
1200	34.0	42.0	47.9
1350	29.6	33.9	40.7

[a] Secant Young's modulus measured during 4-point bend test. Not corrected for porosity.

Seven silicon-densified SiC materials were evaluated on this program. Their strength-temperature behavior is shown in Figure 64. At elevated temperature, their elastic properties are determined by the amount and distribution of the silicon metal phase. This is illustrated in Table 18 and in Figures 65 and 66.

As discussed above, static fatigue is of primary concern in structural ceramics. Figure 67 illustrates the fracture stress vs. time-to-failure from dynamic fatigue tests. The n value in Figure 67 is related to the slope of the strength-time relation, and is the exponent n in the relation $v = AK^n$, where v is the crack velocity and K is the stress intensity at the crack tip (A is a constant). A large value of n means little slow crack growth and time-invariant strength. As shown in Figure 67, SiC exhibits much less SCG and strength degradation than does Si_3N_4.

The mechanisms of SCG and strength degradation appear to be different in Si_3N_4 and SiC. In Si_3N_4 the existence of slow crack growth is usually correlated with a grain boundary sliding mechanism. Accommodation for this deformation is provided by the nucleation of intergranular voids or the extension of pre-existing grain triple-point voids. The result of this is readily observable on Si_3N_4 fracture surfaces.

The current speculation for SiC is that slow crack growth is an atomistic-level process occurring at the crack tip, being related to stress corrosion by an oxidation mechanism. The crack path is generally transgranular in SiC, and SCG is not readily observable on fracture surfaces at the level of magnificaition used on this program. Srinivasan et al.[50] report no detectable SCG in sintered α-SiC tested in argon at 1500°C. However, in air at 1500°C there was evidence of significant SCG, the mechanism being atmospheric attack of intergranular regions leading to grain separation. McHenry and Tressler[51] also indicate the mechanism of SCG in NC-203 HP-SiC to be stress-corrosion by oxidation.

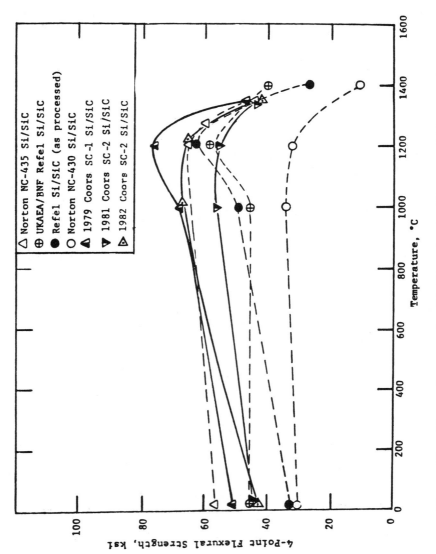

Figure 64. Flexural strength of siliconized silicon carbide materials.

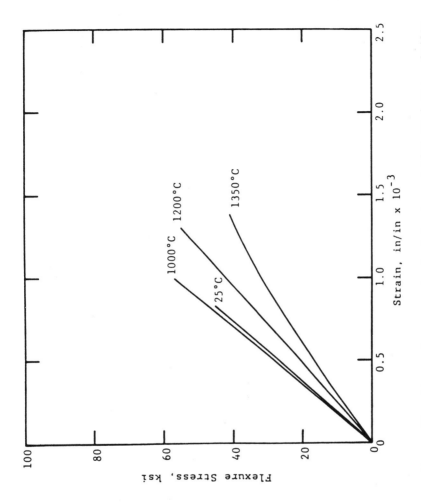

Figure 65. Flexural stress-strain behavior of 1981 SC-2 Coors Si/SiC.

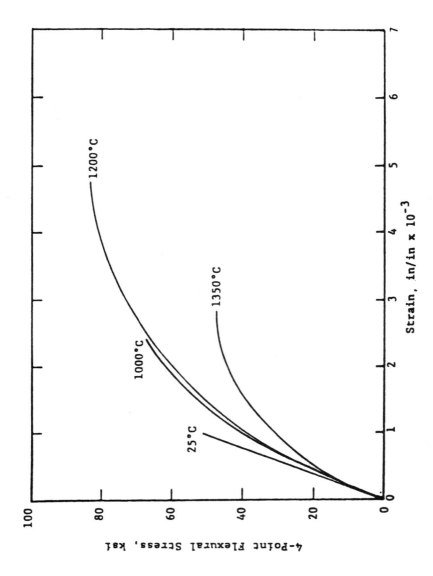

Figure 66. Representative flexural stress-strain behavior of 1979 SC-1 Coors Si/SiC.

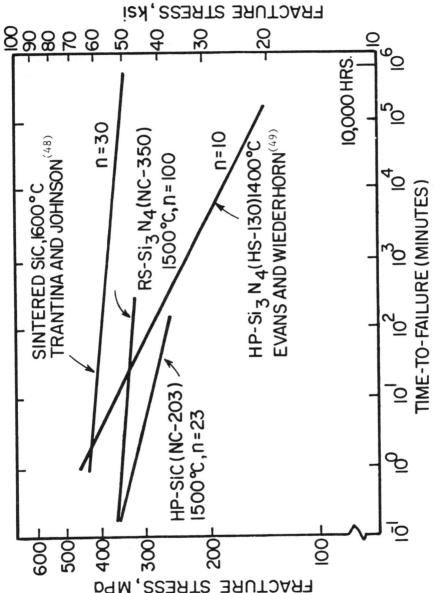

Figure 67. Strength degradation in silicon carbide and silicon nitride due to subcritical crack growth.

7.3 SUMMARY OF HIGH TEMPERATURE FRACTURE ORIGINS IN Si_3N_4 AND SiC

For convenience and reference, the sources of fracture at elevated temperature that have been observed on this program for both Si_3N_4 and SiC materials are tabulated in Table 19. It is noted that all fractography was performed with a low magnification optical binocular microscope.

TABLE 19. SOURCES OF FRACTURE AT ELEVATED TEMPERATURE – Si_3N_4 AND SiC MATERIALS

Material	Source of Fracture[a]
A. Hot-Pressed Si_3N_4	
• Norton NC-132 HP-Si_3N_4 (1% MgO)	Machining and undetermined flaws at 1200°C; some SCG at 1350°C, 50% SCG at 1500°C
• Norton NCX-34 HP-Si_3N_4 (8% Y_2O_3)	Primarily inclusions at 1000°C, surface oxidation and undetermined processing flaws at 1250°C, 50% SCG at 1500°C
• Harbison-Walker HP-Si_3N_4 (10% CeO_2)	Some SCG at 1000°C, prevalent at 1250°C
• Kyocera SN-3 HP-Si_3N_4 (4% MgO, 5% Al_2O_3)	Inclusions at 750°C, less than 10% SCG at 1000°C, ~50% SCG at 1125°C, 100% SCG at 1250°C
• Ceradyne Ceralloy 147A, HP-Si_3N_4 (1% MgO)	Inclusions at 1000°C, ~10% SCG at 1250°C; 50-75% SCG at 1500°C
• Ceradyne Ceralloy 147Y, HP-Si_3N_4 (15% Y_2O_3)	Undetermined processing flaws at 1000°C, less than 10% SCG at 1250°C, 100% SCG at 1500°C
• Ceradyne Ceralloy 147Y-1, HP-Si_3N_4 (8% Y_2O_3)	Inclusions and undetermined processing flaws at 1000°C, less than 5% SCG at 1250°C, 100% SCG at 1500°C
• Fiber Materials HP-Si_3N_4 (4% MgO)	Some SCG at 1000°C; 50% at 1250°C, much at 1400°C
• Toshiba HP-Si_3N_4 (4% Y_2O_3, 3% Al_2O_3)	Dark inclusions at 1000°C, light inclusions at 1250°C; less than 10% SCG at 1350°C, ~50% SCG at 1500°C
• Toshiba HP-Si_3N_4 (3% Y_2O_3, 4% Al_2O_3, SiO_2)	Inclusions at 1000°C, less than 10% SCG at 1125°C and 1250°C, >50% at 1350°C, 100% SCG or plastic deformation at 1500°C
• Westinghouse HP-Si_3N_4 (4% Y_2O_3, SiO_2)	Primarily undetermined at 1000° and 1250°C, 25% SCG at 1500°C
• NASA/AVCO/Norton HP-Si_3N_4 (10% ZrO_2)	Machining flaws at 1000°C, undetermined processing defects at 1250° and 1500°C; no SCG evident at 1500°C
• Battelle HIP-Si_3N_4 (5% Y_2O_3)	SCG less than 5% at 1400°C

TABLE 19 (cont.)

Material	Source of Fracture
B. Reaction-Sintered Si_3N_4	
• 1976 Norton NC-350 RS-Si_3N_4	Pore agglomerates, pores, and undetermined processing defects at 1200°, 1350°, and 1500°C
• Kawecki-Berylco RS-Si_3N_4	Inclusions and pores at 1200° and 1350°C; pores and pore agglomerates, inclusions at 1500°C
• Ford Injection Molded RS-Si_3N_4	Inclusions at 1000°C, inclusions and some pores at 1250° and 1500°C
• AiResearch Slip Cast RS-Si_3N_4 (Airceram RBN-101)	Inclusions and pore clusters at 1000° and 1250°C, pores at 1500°C
• Raytheon Isopressed RS-Si_3N_4	Inclusions at 1000°C, inclusions and pores at 1250° and 1500°C; some oxidation pit origins at 1500°C
• Indussa/Nippon Denko RS-Si_3N_4	Primarily undetermined at 1250° and 1500°C, a few pores as fracture origins
• AiResearch Injection Molded RS-Si_3N_4 (Airceram RBN-122)	Large particles at 1000°, 1250°, and 1500°C; some with associated pores
• 1979 Norton NC-350 RS-Si_3N_4	Half inclusions, half undetermined at 1000°, 1250°, and 1500°C
• Annawerk Ceranox NR-115H RS-Si_3N_4	Primary inclusions, some pores at 1000°C, inclusions at 1250° and 1500°C
• Associated Engineering Developments (AED) Nitrasil RS-Si_3N_4	Even distribution of free Si particles and large pores as origins at 1000°, 1250°, and 1500°C
• Georgia Tech RS-Si_3N_4	Machining damage at 1000°C, pores as 1250°C, undetermined at 1500°C
• AME RS-Si_3N_4	Undetermined at 1400°C
• AiResearch RBN-104 RS-Si_3N_4	Subsurface inclusion at 1400°C

TABLE 19 (cont.)

Material	Source of Fracture

C. Sintered Si_3N_4

- Kyocera SN-205 Sintered Si_3N_4 (5% MgO, 9% Al_2O_3)

  Pores and inclusions at 750°C, less than 10% SCG at 1000°C, ~15% SCG at 1125°C, 25% SCG at 1250°C

- Kyocera SN-201 Sintered Si_3N_4 (4% MgO, 7% Al_2O_3)

  Inclusions at 750°C, less than 5% SCG at 1000°C, SCG more prevalent at 1125° and 1250°C

- GTE Sintered Si_3N_4 (6% Y_2O_3)

  Inclusions at 1200°C, ~50% SCG at 1350°C, 100% SCG at 1500°C

- AiResearch Sintered Si_3N_4 (8% Y_2O_3, 4% Al_2O_3)

  Surface porosity at 1000°C, less than 25% SCG at 1250°C, 75-100% SCG at 1350°C

- Rocketdyne SN-50 Sintered Si_3N_4 (6% Y_2O_3, 4% Al_2O_3)

  Inclusions at 1000°C, ~10% SCG at 1250°C, ~20% SCG at 1283°C, ~30% SCG at 1325°C

- Rocketdyne SN-104 and SN-46 Sintered Si_3N_4 (14% Y_2O_3, 7% SiO_2)

  Inclusions at 1000°C, pores at 1250°C, undetermined at 1325°C

D. Hot-Pressed SiC

- Norton NC-203 HP-SiC (~2% Al_2O_3)

  Undetermined fracture sources at all elevated temperatures

- Ceradyne Ceralloy 146A, HP-SiC (2% Al_2O_3)

  Undetermined at 1000°, 1250°, and 1500°C; possibly some SCG at 1500°C

- Ceradyne Ceralloy 146I, HP-SiC (2% B_4C)

  Primarily undetermined at 1000° and 1250°C, undetermined and possibly some SCG at 1500°C

E. Sintered SiC

- General Electric Sintered β-SiC

  Regions of incomplete sintering as origins at 1000°, 1250°, and 1500°C

- Carborundum Sintered α-SiC (1977 vintage)

  Subsurface pore fracture origins at 1000°, 1250°, and 1500°C

- Kyocera SC-201 Sintered α-SiC (1980 vintage)

  Open and subsurface pores at 1000°C and 1250°C; subsurface pores at 1500°C

TABLE 19 (cont.)

Material	Source of Fracture
• Carborundum 1981 SASC (Hexoloy SX-05)	Subsurface pores and surface finish at 1000° and 1250°C, subsurface pores at 1500°C
• ESK Sintered α-SiC	Subsurface and open porosity at 1000°, 1250°, and 1500°C
F. Silicon-Densified SiC	
• Norton NC-435 Si/SiC	Pores and large grains at 1200°C, undetermined processing flaws at 1275° and 1350°C
• UKAEA/BNF Refel Si/SiC • diamond-ground • as-processed	Surface and subsurface porosity at 1000°, 1200°, and 1400°C Large grains and subsurface pores at 1000°C, undetermined processing flaws at 1200° and 1400°C
• Norton NC-430 Si/SiC	Undetermined at 1000° and 1200°C; undetermined at 1400°C, noticeable plastic deformation
• Coors Si/SiC (1979, SC-1)	Large SiC grains at 1000°C, undetermined at 1200° and 1350°C
• Coors Si/SiC (1981, SC-2)	Surface and subsurface pores at 1000°C, undetermined at 1200° and 1350°C
• Coors Si/SiC (1982, SC-2)	Oxidation pits at 1000° and 1200°C, undetermined at 1350°C

[a] Percentages indicate amount of sample cross-section affected by subcritical crack growth as observed on the fracture surfaces. All fracture surface analysis performed with low magnification optical binocular microscope.

8. Creep Behavior

Creep resistance is of primary concern in the rotating components of a turbine engine. High creep rates can lead to both excessive deformation and uncontrolled stresses.

Creep is a thermally activated process, characterized by a linear or power law stress dependence. After load application there is an instantaneous strain, followed by three regions of creep behavior. Primary creep (transient) is characterized by a creep rate that continually decreases with time. This is followed by secondary (steady-state) creep behavior where the creep rate remains constant with increasing strain and time. Tertiary creep can be observed just prior to fracture in some materials and is characterized by a rapid increase in creep rate. Steady-state secondary creep behavior is of greatest interest in structural applications. The steady-state creep rate, $\dot{\varepsilon}$, is governed by an empirical relation, the general form of which is:

$$\dot{\varepsilon} = A\sigma^n e^{-E/kT}$$

Creep deformation mechanisms are deduced in part by analysis of measured creep rates with respect to this relation. Experimental schedules normally cover a range of stress, σ, and temperature, T, which permit analytical determination of the stress exponent, n, and activation energy, E. The term A in the empirical relation above is a constant. A knowledge of the stress exponent and activation energy permits some mechanistic interpretation, which is important in understanding the measured behavior and eventually improving materials properties. However, this information alone is not sufficient for a comprehensive mechanistic study. Microstructural analysis of crept specimens using light microscopy and scanning, replica, and transmission electron microscopy can reveal features that suggest the rate-controlling mechanisms involved. Such features include grain boundary sliding, nucleation of microcracks, porosity on grain boundaries, dislocations, diffusion within grains, diffusion along grain boundaries, etc.

The creep data obtained on this program and its predecessor for various HP-Si_3N_4 materials are presented in Figure 68, where the steady-state creep strain rate is plotted as a function of applied stress. The stress dependence and activation energy data are tabulated in Table 20. Figure 69 and Table 21 contain similar information for the reaction-sintered silicon nitride materials evaluated. The data for SiC are presented in Figure 70 and Table 22.

This information is summarized in Figure 71, where data bands of creep behavior for silicon nitride and silicon carbide materials are plotted as a function of processing method.

8.1 HP-Si_3N_4

For hot-pressed Si_3N_4, good results are beginning to be achieved in recent processing efforts to control the composition and thus properties of the grain boundary phases. Recall that MgO-doped HPSN had an amorphous intergranular phase that was not resistant to deformation under stress. Grain boundary plasticity and associated cavity nucleation and growth resulted in low high-temperature strengths and high creep rates. The high degree of stress dependence of the creep rate in such MgO-modified HP-Si_3N_4, characterized by $1.5 < n < 1.8$, suggested such a viscoelastic mechanism.

Efforts were then directed toward Y_2O_3-modified Si_3N_4 structures. At 1350°C, HP-Si_3N_4 materials by Ceradyne (8% Y_2O_3) and Toshiba (3% Y_2O_3, 4% Al_2O_3, SiO_2) did not exhibit improved creep resistance when compared to magnesia-doped Norton NC-132. This is illustrated in Figure 71, where curve A falls within the band of data for MgO-doped HP-Si_3N_4. We then evaluated two other Y_2O_3-doped materials at 1350°C, Norton NCX-34 (8% Y_2O_3) and Toshiba (4% Y_2O_3, 3% Al_2O_3). The results for these two materials are represented by curve B in Figure 71. It is observed that some

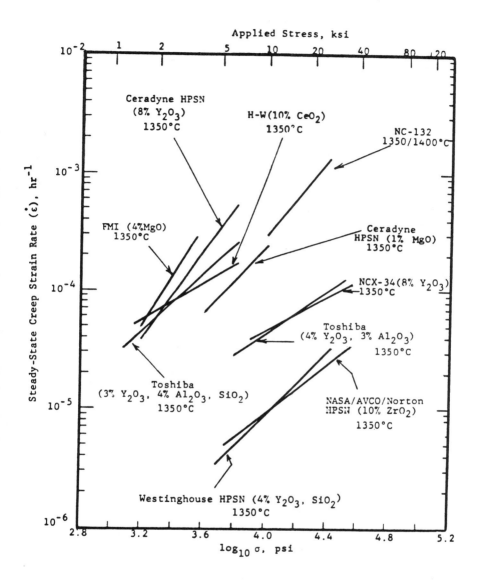

Figure 68. Steady-state flexural creep rate vs. applied stress for various hot-pressed Si_3N_4 materials at 1350°C.

TABLE 20. DERIVED CREEP STRESS DEPENDENCE AND ACTIVATION ENERGY FOR VARIOUS Si_3N_4 MATERIALS EVALUATED.

Material	Temp., °C	Stress Exponent, n	Activation Energy, kcal/mole
Hot-Pressed			
Norton NC-132 (1% MgO)	1350 / 1500	1.6	130[a]
Ceradyne Ceralloy 147A (1% MgO)	1350	1.5	--
Norton NCX-34 (8% Y_2O_3)	1350 / 1500	0.7-0.8	91
Ceradyne Ceralloy 147Y-1 (8% Y_2O_3)	1350	1.8	--
Ceradyne Ceralloy 147Y (15% Y_2O_3)	1350	0.9[b]	--
Fiber Materials (4% MgO)	1250 / 1350	~0.5 / 2.3	--
Harbison-Walker (10% CeO_2)	1350	0.9	--
Toshiba (4% Y_2O_3, 3% Al_2O_3)	1350	0.9	--
Toshiba (3% Y_2O_3, 4% Al_2O_3, SiO_2)	1350	1.4	--
Westinghouse (4% Y_2O_3, SiO_2)	1350	1.3	--
NASA/AVCO/Norton (10% ZrO_2)	1350	1.0	--
Sintered			
Rocketdyne SN-46, SN-104 (14% Y_2O_3, 7% SiO_2)	1250 / 1350	1.1 / --	114

[a] Computed from IITRI and Seltzer [52] data.
[b] Material decomposed, exhibited spontaneous catastrophic phase instability.

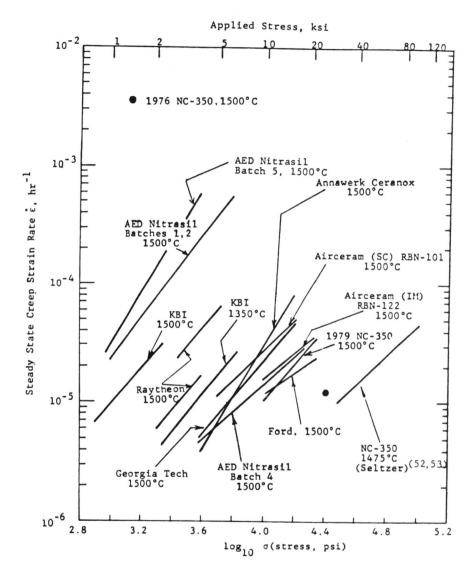

Figure 69. Steady-state flexural creep rate vs. applied stress for various reaction-sintered Si_3N_4 materials.

TABLE 21. DERIVED CREEP STRESS DEPENDENCE AND ACTIVATION ENERGY FOR VARIOUS RS-Si_3N_4 MATERIALS STUDIED

Material	Density g cm^{-3}	Temp., °C	Stress Exponent, n	Activation Energy, kcal/mole
Norton NC-350 (1976)	2.41-2.55	1500	1.3[a]	~90[b]
KBI	2.35-2.53	1350 1500	1.5-1.6	74
Ford (IM)	2.76	1500	0.9	--
AiResearch (IM) Airceram RBN-122	2.68	1500	1.1	--
AiResearch (SC) Airceram	2.88	1500	1.3	--
Norton NC-350 (1979)	2.34-2.47	1500	1.7	--
Annawerk Ceranox NR-115H	2.65	1500	2.2	--
AED Nitrasil				
Batch 1,2	2.52	1500	1.7	--
Batch 4	2.61	1500	1.2	--
Batch 5	2.63	1500	2.1	--
Raytheon (IP) RSSN	2.37-2.42	1500	1.7	--
Georgia Tech	2.49-2.64	1500	1.6	--

[a] Data of Seltzer.[52,53]

[b] Computed from IITRI and Seltzer data.

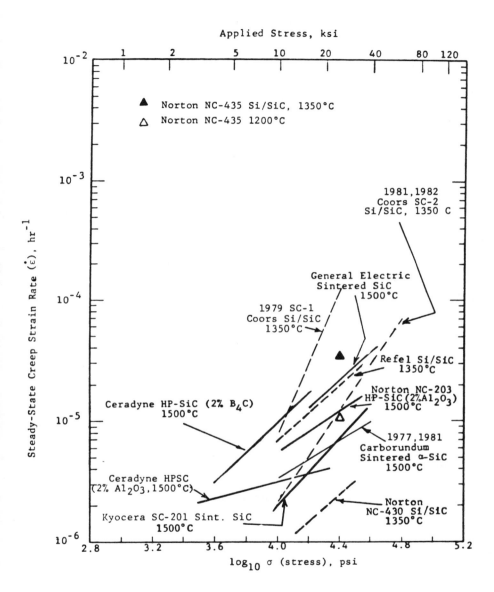

Figure 70. Steady-state flexural creep rate vs. applied stress for various silicon carbide materials.

TABLE 22. DERIVED CREEP STRESS DEPENDENCE AND ACTIVATION ENERGY FOR VARIOUS SiC MATERIALS STUDIED

Material	Temp., °C	Stress Exponent, n	Activation Energy, kcal/mole
Sintered SiC			
General Electric Sintered SiC	1500	1.1-1.2	210-250[a]
1977, 1981 Carborundum Sintered SiC	1500	0.8-1.1	--
Kyocera SC-201 Sintered SiC	1500	1.4	--
Hot-Pressed SiC			
Ceradyne Ceralloy 146A HP-SiC (2% Al_2O_3)	1500	0.3	--
Ceradyne Ceralloy 146I HP-SiC (2% B_4C)	1500	1.5	--
Norton NC-203 HP-SiC (2% Al_2O_3)	1500	0.9	--
Silicon-Densified SiC			
Norton NC-435 Si/SiC	1200 1350	0.9[b]	235[c]
UKAEA/BNF Refel Si/SiC	1350	1.2	--
Norton NC-430 Si/SiC	1350	1.2	--
1979 Coors SC-1 Si/SiC	1350	2.9	--
1981, 1982 Coors SC-2 Si/SiC	1350	1.9	--

[a] General Electric data.[54]
[b] Data of Seltzer.[52,53]
[c] Computed from IITRI and Seltzer data.

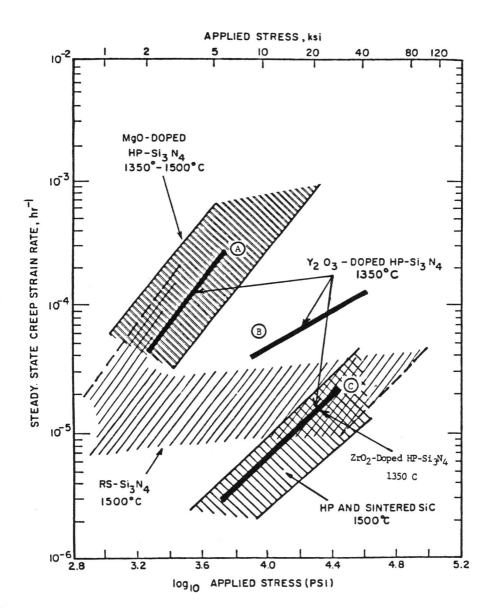

Figure 71. Flexural creep behavior of various Si_3N_4 and SiC materials.

improvement in creep resistance was beginning to be attained. This was accomplished by the controlled crystallization of the intergranular phase in these materials.

Our most recent results for hot-pressed Si_3N_4 are quite encouraging. Much improved creep resistance was observed for the Westinghouse material, doped with 4% Y_2O_3 and an undetermined amount of SiO_2. This is illustrated in Figure 71 by curve C (also shown in Figure 68). The same is true for the NASA/AVCO/Norton material, which is prepared with nominally 10% ZrO_2 additive (also represented by curve C in Figure 71). Thus, Y_2O_3 and ZrO_2 dopants for Si_3N_4 are beginning to fulfill their promise of improved properties. Note that these two materials fall into the data band for SiC materials in Figure 71. Additionally, low stress-dependence is observed (1.0 < n < 1.3, Table 20). These are the best HP-Si_3N_4 materials investigated on this program. These creep results correlate with the good high temperature strength reported above.

Recall that the NASA/AVCO/Norton HP-Si_3N_4 (10% ZrO_2) material exhibited excellent long-term elevated temperature behavior in that a minimal amount of static fatigue was observed. The low creep rate and stress dependence obtained for the material is consistent with this, and is also a result of the deformation-resistant grain boundary phase. The Naval Research Laboratory has reported initial results in attempts to obtain more refractory silicate intergranular phases in HP-Si_3N_4 with zirconium-based densification additives such as ZrO_2, $ZrSiO_4$, ZrC, and ZrN.[55-57] We have evaluated early versions of ZrO_2-doped NRL HP-Si_3N_4 on the previous AFML program.[58] The concept here is that zircon ($ZrSiO_4$) would be present at the grain boundaries either by the direct use of it as an additive, or through the reaction of the additive binary zirconium compounds with surface-absorbed SiO_2. Rice and McDonough[55] have pointed out that zirconium silicate is a desirable intergranular compound since it has a reported melting point as high as 2500°C and solidus temperature of >1775°C (compared to >1500°C for magnesium silicate and

>1300°C for magnesium silicate with a few wt% CaO impurity). Furthermore, the Young's elastic modulus and coefficient of thermal expansion of $ZrSiO_4$ (~25 x 10^6 psi and 5 x 10^{-6} °C^{-1}, respectively) are generally better matched to those of Si_3N_4 (~45 x 10^6 psi and 3 x 10^{-6} °C^{-1}, respectively) than are those of the lower modulus, higher expansion magnesium silicates (~20 x 10^6 psi and ~11 x 10^{-6} °C^{-1}, respectively).[55]

In summary, Y_2O_3 and ZrO_2 dopants for HP-Si_3N_4 are beginning to fulfill their promise of improved properties. The creep results for the Westinghouse and NASA/AVCO/Norton materials fall within the band of behavior we have previously reported for the very creep-resistant SiC materials (Figure 71). This is a significant advancement in Si_3N_4 technology.

8.2 RS-Si_3N_4 MATERIALS

The RS-Si_3N_4 materials studied to date generally have very good creep strengths due to the lack of significant oxide impurity phases. Stress exponents are lower than for HP-Si_3N_4, being typically 1.1 ± 0.2. Grain boundary sliding models have been proposed for reaction sintered Si_3N_4,[59,60] where accommodation is by microcracking and boundary separation. However, other investigators contend that the rate-controlling process in Si_3N_4 is not the grain boundary sliding itself, but rather the microcrack formation necessary to accommodate the relative crystal movement.[61-63]

8.3 SiC MATERIALS

For all forms of SiC studied, especially hot-pressed and sintered, creep rates are extremely low, and a linear stress dependence is observed. This is observed in Figure 70 and Table 22. Siliconized SiC materials have substantially higher creep rates, probably due to the presence of the continuous silicon metal phase. This phase is very near its melting point under the creep test conditions, and would be expected to exhibit high rates of creep deformation. Since the creep strength of hot-pressed and sintered SiC is so high (i.e., creep rate so low) very few applications studies have identified creep deformation to be a predominant failure mode for SiC. For this reason, only a few mechanistic studies have been undertaken. The linear stress dependence of the creep rate in SiC suggests that diffusion is the rate-controlling process. A carbon-vacancy diffusion mechanism has been proposed.[64,65]

9. Thermal Expansion

One of the reasons that silicon-base ceramics are prime candidates for use in advanced gas turbine applications is their low expansion coefficient and high strength, which makes them less susceptible than many other ceramics (especially oxides) to thermal shock damage. However, it is recognized that such low thermal expansion can be a disadvantage also; for instance, in the creation of thermal expansion mismatch situations with higher expansion metal engine components. As with many other aspects of heat engine materials selection and component design, many trade-offs exist.

Thermal expansion is perhaps the least variable property of silicon ceramics. It is mainly a function of the solid phase and thus not strongly affected by porosity and minor impurities. Figure 72 presents thermal expansion data bands for various silicon-based ceramics. Details for specific materials are contained in Figure 73. Tabular data for the various forms of silicon nitride and silicon carbide investigated on this program are presented in Tables 23 and 24, respectively.

All forms of SiC have ~50% higher thermal expansion than all forms of Si_3N_4. Within a given material performance band, the effect of additives can be seen. For instance, within the thermal expansion data band for SiC shown in Figure 72 exist curves for additive-free sintered SiC, hot-pressed SiC (which contains ~2% Al_2O_3), and siliconized SiC (which contains ~10-20% silicon metal). Siliconized forms of SiC exhibit the lowest expansion, sintered SiC has intermediate expansion, and Al_2O_3-doped hot-pressed SiC the highest expansion. The low expansion for Si/SiC is due to the low expansion of the silicon metal phase. The high expansion for hot-pressed SiC is caused by their slight alumino-silicate grain boundary phase (oxides having much higher expansion than pure SiC).

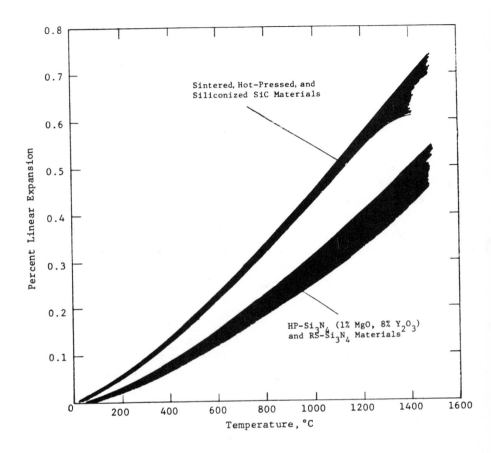

Figure 72. Thermal expansion data bands for Si_3N_4 and SiC materials.

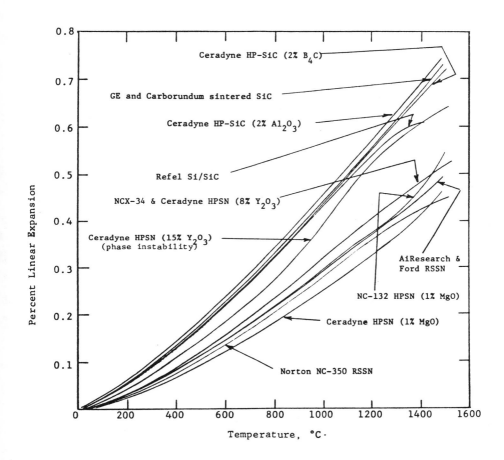

Figure 73. Thermal expansion of various Si_3N_4 and SiC materials.

TABLE 23. THERMAL EXPANSION OF Si_3N_4 MATERIALS

Material	Percent Linear Expansion at 1000°C	Mean Coefficient of Thermal Expansion, (20°-1000°C), $10^{-6}/°C$
Hot-Pressed Si_3N_4		
Norton NC-132 (1% MgO)	.295	3.01
Ceradyne 147A (1% MgO)	.299	3.05
Fiber Materials (4% MgO)	.336	3.43
Norton NCX-34 (8% Y_2O_3)	.315	3.21
Ceradyne 147Y-1 (8% Y_2O_3)	.322	3.29
Ceradyne 147Y (15% Y_2O_3)	.385	3.93
Harbison-Walker (10% CeO_2)	.341	3.48
Toshiba (4% Y_2O_3, 3% Al_2O_3)	.305	3.11
Toshiba (3% Y_2O_3, 4% Al_2O_3, SiO_2)	.316	3.22
NASA/AVCO/Norton (10% ZrO_2)	.355	3.62
Reaction-Sintered Si_3N_4		
Norton NC-350 (1976)	.274	2.80
Kawecki-Berylco	.280	2.86
Ford (IM)	.300	3.06
AiResearch (SC) RBN-101	.297	3.03
AiResearch (IM) RBN-122	.296	3.02
Raytheon	.289	2.95
Norton NC-350 (1979)	.292	2.98
AED Nitrasil (Batch 4)	.302	3.08
Georgia Tech	.268	2.73

TABLE 24. THERMAL EXPANSION OF SiC MATERIALS

Material	Percent Linear Expansion at 1000°C	Mean Coefficient of Thermal Expansion (20°-1000°C), 10^{-6}/°C
Hot Pressed SiC		
Ceradyne 146A (2% Al_2O_3)	.449	4.58
Ceradyne 146I (2% B_4C)	.438	4.47
Norton NC-203 (2% Al_2O_3)	.451	4.60
Sintered SiC		
Carborundum 1977 α-SiC	.438	4.47
General Electric β-SiC	.432	4.41
Kyocera SC-201 α-SiC	.445	4.54
Carborundum 1981 α-SiC (Hexoloy SX-05)	.430	4.39
Siliconized SiC		
Norton NC-435	.427	4.36
UKAEA/BNF Refel	.424	4.33
Norton NC-430	.424	4.33
Coors 1979 (SC-1)	.425	4.34
Coors 1981 (SC-2)	.420	4.29

Trends for Si_3N_4 materials are stronger and directly correlated with the amount of intergranular oxide phase present in the microstructure. This is seen by consideration of the tabular data in Table 23. In the limit of minimal intergranular phases, the lowest thermal expansion is observed. This represents the pure $RS-Si_3N_4$ materials. The hot-pressed forms of Si_3N_4 all contain varying amounts of the high expansion oxide phase in intergranular regions.

Compared to the various forms of Si_3N_4, the thermal expansion of SiC only varies a maximum of ~5% as a function of processing method. As shown in Table 23, the difference in thermal expansion between a reaction-sintered Si_3N_4 and a highly doped hot-pressed Si_3N_4 can be as much as 30%. As mentioned previously, this is another indication that the properties of the various forms of SiC are more intrinsic, and that the properties of Si_3N_4 are mainly dependent on the existence or the lack of an oxide grain boundary phase, and its specific character (i.e., phase composition, structure, and resulting mechanical properties).

10. Thermal Shock Resistance

The ability to withstand the thermal stresses generated during ignition, flame-out, and operating temperature excursions is an important consideration in evaluating potential ceramic heat engine materials. Thermally created stresses may initiate a fracture which can result in a catastrophic failure, or cause existing flaws to grow giving a gradual loss of strength and eventual loss of component integrity. However, the evaluation of thermal stress resistance is a complex task since performance is dependent not only on material thermal and mechanical properties, but is also influenced by heat transfer and geometric factors (i.e., heat transfer coefficient and component size).

Thermal shock resistance was determined on this program by the water quench method, with the initiation of thermal shock damage being detected by internal friction measurement. The methodology for this was discussed in detail in Section 5.5. In conducting this test, internal friction is measured before and after water quench from successively higher temperatures using the flexural resonant frequency Zener bandwidth method. A marked change in internal friction (specific damping capacity) indicates the onset of thermal shock damage (i.e., thermal stress-induced crack initiation). This defines the critical quench temperature difference, ΔT_c, which is compared to analytical thermal stress resistance parameters. The parameter which is most applicable to the experimental severe water quench is $R = \sigma(1 - \mu)/\alpha E$,[66] where σ is the strength, μ is Poisson's ratio, α is thermal expansion, and E is the elastic modulus. The details, interpretation, and limitations of this procedure for assessing relative thermal shock resistance was discussed previously.[67]

Hasselman's theory[68] unified considerations of thermal stress fracture initiation and the degree of damage due to subsequent crack propagation. It was demonstrated that the material parameters that are required to give greatest resistance to thermal stress fracture initiation (high strength, low Young's elastic modulus, high thermal conductivity, low thermal expansion, etc.) are, in general, mutually exclusive to the requirements providing greatest resistance to crack propagation and material damage due to thermal stresses (low strength, high modulus, high fracture surface energy, etc.). These properties and the behavior of a material experiencing thermally created stresses are influenced by the initial size and density of cracks or flaws (i.e., the total flaw structure of the material). This structure affects the material's ability to store elastic thermal strain energy, the driving force for crack propagation. In general a material resistant to fracture has a low flaw density, so once a sufficiently high stress level (or level of stress intensity) is obtained in such a material, its high stored energy causes the most deleterious flaw to propagate catastrophically or in a kinetic manner. This occurs because the energy release rate is greater than that needed to merely balance fracture surface energy. Resistance to thermal stress damage means the material is resistant to crack propagation, as opposed to fracture initiation. When cracks are propagated due to thermal stresses in such a case, the propagation is quasi-static in that only enough crack length is generated to absorb the available strain energy. The rate of strain energy release is lower in such materials and less strain energy is available as a driving force for continued propagation. This usually occurs in a more highly flawed material. This discussion shows why internal friction, as a measure of the total flaw spectrum, is useful in monitoring thermal shock performance.

The baseline (unshocked) levels of internal friction for all silicon-base materials evaluated on this program are presented in Tables 25-27. Figure 74 compares the observed internal friction-quench temperature difference behavior for various general classes of silicon ceramics. Table 28 presents the various analytical thermal stress resistance parameters that have been

TABLE 25. BASELINE UNSHOCKED LEVEL OF INTERNAL FRICTION FOR HOT PRESSED Si_3N_4 MATERIALS

	Material	Baseline Internal Friction Approximate Range, x 10^4
1.	Norton NC-132 HPSN (1% MgO)	1.0
2.	Norton NCX-34 HPSN (8% Y_2O_3)	0.5-0.8
3.	Ceradyne 147A HPSN (1% MgO)	0.9-2.7
4.	Ceradyne 147-Y-1 HPSN (8% Y_2O_3)	0.9-1.2
5.	Ceradyne 147-Y HPSN (15% Y_2O_3)	0.7-1.3
6.	Fiber Materials HPSN (4% MgO)	1.1-1.4
7.	Harbison-Walker HPSN (10% CeO_2)	0.6
8.	Toshiba HPSN (4% Y_2O_3, 3% Al_2O_3)	0.9
9.	Toshiba HPSN (3% Y_2O_3, 4% Al_2O_3, SiO_2)	0.8-1.4
10.	Westinghouse HPSN (4% Y_2O_3, SiO_2)	0.6-1.0
11.	NASA/AVCO/Norton HPSN (10% ZrO_2)	3.5

TABLE 26. BASELINE UNSHOCKED LEVELS OF INTERNAL
FRICTION FOR REACTION SINTERED Si_3N_4 MATERIALS

	Material	Baseline Internal Friction Approximate Range, $\times 10^4$
1.	Norton NC-350 RSSN (1976)	2-3
2.	Kawecki-Berylco RSSN	2-5
3.	Ford (IM) RSSN	1.7-2.5
4.	AiResearch (SC) Airceram RBN-101	4
5.	AiResearch (IM) Airceram RBN-122	1.4-2.6
6.	Raytheon (IP) RSSN	2.8-4.9
7.	Norton NC-350 RSSN (1979)	2.2-2.7
8.	AED Nitrasil RSSN	1.4-2.2
9.	Georgia Tech RSSN	1.9-2.2

TABLE 27. BASELINE INTERNAL FRICTION OF SiC MATERIALS

	Material	Baseline Internal Friction Approximate Range, $\times 10^4$
1.	Norton NC-435 Si/SiC	1-2
2.	Carborundum Sintered α-SiC	1.6-2.3
3.	General Electric Sintered β-SiC	2-2.7
4.	UKAEA/BNF Refel Si/SiC	0.6-2.7
5.	UKAEA/BNF Refel Si/SiC	5-10
6.	Ceradyne 146A HPSC (2% Al_2O_3)	1.4-2.4
7.	Ceradyne 146I HPSC (2% B_4C)	1.2-3.7
8.	Norton NC-203 HPSC (2% Al_2O_3)	1.2
9.	Norton NC-430 Si/SiC	1-1.5
10.	Coors Si/SiC (1979, SC-1)	0.9
11.	Kyocera SC-201 Sintered SiC	2.7-3.3
12.	1981 Coors SC-2 Si/SiC	1.5

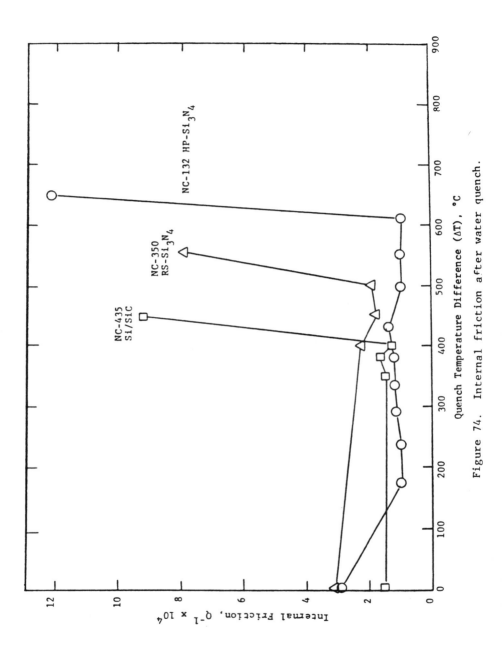

Figure 74. Internal friction after water quench.

TABLE 28. THERMAL STRESS RESISTANCE PARAMETERS(66)

LITERATURE DESIGNATION	PARAMETER TYPE	PARAMETERS	PHYSICAL INTERPRETATION/ENVIRONMENT	TYPICAL UNITS
R	Resistance to Fracture Initiation	$\dfrac{\sigma_t(1-\mu)}{\alpha E}$	Maximum ΔT allowable for steady heat flow;	°C
R'	Resistance to Fracture Initiation	$\dfrac{\sigma_t(1-\mu)k}{\alpha E}$	Maximum heat flux for steady flow;	cal cm^{-1} sec^{-1}
R''	Resistance to Fracture Initiation	$\dfrac{\sigma_t(1-\mu)\alpha_{TH}}{\alpha E}$	Maximum allowable rate of surface heating;	cm^2 °C sec^{-1}
R'''	Resistance to Propagation Damage	$\dfrac{E}{\sigma^2(1-\mu)}$	Minimum in elastic energy at fracture available for crack propagation	(psi)$^{-1}$
R''''	Resistance to Propagation Damage	$\dfrac{YE}{\sigma^2(1-\mu)}$	Minimum in extent of crack propagation on initiation of thermal stress fracture	cm
R_{st}	Resistance to further crack propagation	$\left[\dfrac{Y}{\alpha^2 E}\right]^{1/2}$	Minimum ΔT allowed for propagation long cracks	°C m$^{-1/2}$

σ_t = Tensile Strength
μ = Poisson's ratio
α = Coefficient of Thermal Expansion
E = Young's Modulus of Elasticity

ρ = Density
C_p = Specific Heat
k = Thermal Conductivity
α_{TH} = Thermal Diffusivity
Y = Fracture Surface Energy

derived. For the water-quench test, it was mentioned above that the R parameter is most appropriate.

Cumulative analytical and experimental thermal shock results are compiled in Tables 29 through 31 for hot-pressed and reaction-sintered Si_3N_4, and SiC materials, respectively. The pertinent thermal and mechanical properties used to compute R, the analytical thermal stress resistance parameter, are also shown. The computed R value for each material represents the maximum ΔT allowable before shock initiation. The R parameter is compared with experimentally obtained ΔT_c values from the water quench tests, as shown. Figure 75 provides a direct comparison of the analytical and experimental thermal shock results for all materials investigated to date. The trend line is a linear least squares regression data fit for all materials. The materials evaluated follow a definite trend of thermal shock resistance as a function of material type and processing method as shown in Figure 75. That is, in descending order of resistance to damage by thermal shock, the materials are listed: (1) Y_2O_3-modified HP-Si_3N_4, (2) MgO-doped HP-Si_3N_4, (3) high density RS-Si_3N_4, (4) low density RS-Si_3N_4, and finally, (5) all forms of SiC.

A notable aspect of the results illustrated in Figure 75 is the relative thermal shock resistance of the 1976 and 1979 versions of Norton NC-350 RS-Si_3N_4. By virtue of its ~30% higher strength and ~10% lower expansion-modulus product, 1979 NC-350 has significantly increased thermal shock resistance. The more recent Norton NC-350 is the most thermal shock resistant reaction sintered Si_3N_4 evaluated on this program. Figure 75 illustrates that it is better than some of the denser hot-pressed Si_3N_4 materials.

The results for the Toshiba and Westinghouse HP-Si_3N_4 materials are also encouraging. As shown in Figure 75, Toshiba HP-Si_3N_4 (4% Y_2O_3, 3% Al_2O_3) and Westinghouse HP-Si_3N_4 (4% Y_2O_3, SiO_2) exhibited the highest experimentally determined critical quench temperature difference, ΔT_c, obtained to date, i.e., ΔT_c = 725°C and 750°C, respectively. These results for the Toshiba and

TABLE 29. THERMAL STRESS RESISTANCE PARAMETERS FOR HOT-PRESSED Si_3N_4 MATERIALS

Material	Flexural Strength,[a] ksi	Poisson's Ratio[b]	Thermal Expansion,[c] $10^{-6}/°C$	Elastic Modulus, 10^6 psi	$R = \sigma(1-\mu)/\alpha E$,[d] °C	ΔT_c (Δq^{-1}),[e] °C
Norton NC-132 HPSN (1% MgO)	103.1	0.27	2.50	47.1	639	600
Norton NCX-34 HPSN (8% Y_2O_3)	126.7	0.27	2.63	48.6	724	600-800
Ceradyne 147A HPSN (1% MgO)	87.1	0.27	2.60	47.9	511	500-600
Ceradyne 147Y-1 HPSN (8% Y_2O_3)	83.3	0.29	2.61	45.4	499	500-700
Ceradyne 147Y HPSN (15% Y_2O_3)	87.6	0.28	3.0	44.7	470	400-500
Fiber Materials HPSN (4% MgO)	66.8	0.28	2.94	46.2	354	600
Harbison-Walker HPSN (10% CeO_2)	87.9	0.28	2.88	45.0	488	400-600
Toshiba (4% Y_2O_3, 3% Al_2O_3)	105.7	0.28 (est)	2.40	47.5	668	725
Toshiba (3% Y_2O_3, 4% Al_2O_3, SiO_2)	83.6	0.28 (est)	2.35	41.3	620	625
Westinghouse (4% Y_2O_3, SiO_2)	91.0	0.28 (est)	2.4 (est)	44.2	618	750
NASA/AVCO/Norton HPSN (10% ZrO_2)	95.2	0.28 (est)	2.99	45.0	509	600-650

[a] All properties at 25°C, unless otherwise noted; computed average of all samples tested; note that 4-point bend strength is used instead of tensile strength.
[b] Determine at AFML by resonant sphere technique, unless otherwise indicated.
[c] At 500°C.
[d] Analytical thermal stress resistance parameter; maximum ΔT allowable for steady heat flow.
[e] Critical ΔT determined by change in internal friction in water quench tests.

TABLE 30. THERMAL STRESS RESISTANCE PARAMETERS FOR REACTION-SINTERED Si_3N_4 MATERIALS

Material	Flexural Strength,[a] ksi	Poisson's Ratio[b]	Thermal Expansion,[c] 10^{-6}/°C	Elastic Modulus, 10^6 psi	$R = \sigma(1-\mu)/\alpha E$,[d] °C	ΔT_c (Δq^{-1}),[e] °C
Norton NC-350 RSSN (1976)	29.5	0.22	2.29	25.6	393	400-500
Kawecki-Berylco RSSN	21.1	0.24	2.30	20.9	334	600-800
Ford (IM) RSSN	38.2	0.25	2.48	30.7	376	350-450
AiResearch (SC) Airceram RBN-101	37.9	0.24	2.42	32.0	372	400-600
AiResearch (IM) Airceram RBN-122	32.6	0.23	2.44	30.2	341	350-400
Raytheon (IP) RSSN	21.6	0.24(est.)	2.41	23.8	286	~400
Norton NC-350 RSSN (1979)	37.6	0.22(est.)	2.46	22.1	539	500
AED Nitrasil RSSN	29.1	0.24(est.)	2.67	26.6	311	425-475
Georgia Tech RSSN	26.2	0.24(est.)	2.00	23.9	417	500

[a] All properties at 25°C, unless otherwise noted; computed average of all samples tested; note that 4-point bend strength is used instead of tensile strength.
[b] Determine at AFML by resonant sphere technique, unless otherwise indicated.
[c] At 500°C.
[d] Analytical thermal stress resistance parameter; maximum ΔT allowable for steady heat flow.
[e] Critical ΔT determined by change in internal friction in water quench tests.

TABLE 31. THERMAL STRESS RESISTANCE PARAMETERS FOR SILICON CARBIDE MATERIALS

Material	Flexural Strength,[a] ksi	Poisson's Ratio[b]	Thermal Expansion,[c] $10^{-6}/°C$	Elastic Modulus, 10^6 psi	$R = \sigma(1-\mu)/\alpha E,$[d] °C	ΔT_c (ΔQ^{-1}),[e] °C
Norton NC-435 Si/SiC	57.2	0.17	3.83	50.7	244	350–400
Carborundum Sint. α-SiC	44.2	0.16	3.8	58.2	168	300
General Electric Sint. β-SiC	63.8	0.16	3.78	54.8	259	375–400
UKAEA/BNF Refel Si/SiC	44.9	0.17	3.79	57.5	171	300–400
UKAEA/BNF Refel Si/SiC (as processed)	33.6	0.17	3.79	52.5	140	---
Ceradyne 146A,HP-SiC (2% Al_2O_3)	60.1	0.16	4.0	67.2	188	~300 (est.)
Ceradyne 146I HP-SiC (2% B_4C)	45.6	0.16 (est.)	3.89	65.3	151	300
Norton NC-203 HP-SiC (2% Al_2O_3)	101.8	0.17	4.08	64.7	320	350
Norton NC-430 Si/SiC	30.4	0.17	3.85	58.7	112	300–350
Coors Si/SiC (1979 SC-1)	50.6	0.17 (est.)	3.50	51.1	235	400
Kyocera SC-201 Sint. SiC	56.1	0.17 (est.)	3.70	60.5	208	325
1981 Coors SC-2 Si/SiC	45.4	0.17 (est.)	3.95	56	170	375–400

[a] All properties at 25°C, unless otherwise noted; computed average of all samples tested; note that 4-point bend strength is used instead of tensile strength.
[b] Determine at AFML by resonant sphere technique, unless otherwise indicated.
[c] At 500°C.
[d] Analytical thermal stress resistance parameter; maximum ΔT allowable for steady heat flow.
[e] Critical ΔT determined by change in internal friction in water quench tests.

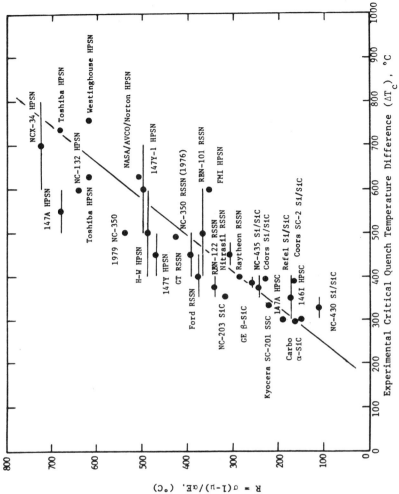

Figure 75. Analytical vs. experimental thermal shock results for various Si_3N_4 and SiC materials.

Westinghouse materials correlate with the high strength of these materials.

To summarize the results for SiC, it is noted that all forms of silicon carbide fall at the opposite extreme of thermal shock resistance when compared to HP-Si_3N_4. This is due to the extremely high thermal expansion and elastic modulus for SiC. It is interesting to note that despite the very high strength of Norton NC-203 HP-SiC (twice as strong as other SiC materials), its thermal shock resistance remains about the same as the other SiC materials. The reason for this is the overriding influence of the high αE product for SiC. For example, for NC-203 to exhibit an R value of 500°C (that is, approximately mid-range of the materials shown in Figure 75), it would have to have a strength of 160 ksi. For NC-203 HP-SiC to have an R value as high as the best HP-Si_3N_4, i.e., R = 750°C, it would have to have a strength of 240 ksi. Thus, SiC inherently has lower thermal shock resistance than Si_3N_4, and no conventional processing methods for SiC can overcome the high αE factor. In certain situations the higher thermal conductivity for SiC helps, but not enough to override the effect of high elastic modulus and high thermal expansion. It is for this reason that SiC has found application in combustors, where high thermal conductivity is useful to minimize hot spots and the temperature is relatively static.

11. Oxidation Behavior

Oxidation of Si_3N_4 and SiC materials is strongly dependent on porosity for reaction sintered materials and specific impurities for hot-pressed materials. In this screening program test bars were exposed in static air for 100, 1000, and occassionally 2000 hr at specific temperatures, as listed in Table 32. The choice of exposure temperature was generally based on the projected upper use temperature for the material. Yttria- and ceria-doped HP-Si_3N_4 materials, however, were all exposed at 1000°C, since a destructive oxidation-related phase instability has sometimes been observed for such materials at intermediate temperatures. Post-exposure evaluation of materials is outlined in Table 33, and includes optical and SEM examination of the oxide scale morphology, weight change, retained strength and fracture source identification (with comparison to unexposed data), and identification of the oxide scale products by X-ray diffraction (XRD) and X-ray fluorescence (XRF) techniques.

The details of the weight change data for specific HP-Si_3N_4, sintered Si_3N_4, RS-Si_3N_4, and SiC materials are presented in Tables 34 through 37, respectively. Similarly, the residual strength and fracture origins for these materials are presented in Tables 38 through 41. The unexposed bend strength and fracture origins are also tabulated for each material for comparison. For convenience, the weight change data are combined and presented in summary form in Table 42. Note that when interpreting weight change data, it is important to recognize that a negligible change in weight does not necessarily mean that the environment was benign. Positive and negative weight changes due to competing mechanisms could conceivably occur at the same rate. The residual strength data are combined and summarized in Table 43. These data for each material can be discussed in terms of the purity and porosity of each material, the optical and SEM surface scale morphology, and the mechanistic interpretation for

TABLE 32. SCHEDULE FOR 100 AND 1000 HOUR STATIC AIR EXPOSURE

Material	Temperature, °C
HP-Si_3N_4 (MgO)	1200
HP-Si_3N_4 (Y_2O_3)	1000
HP-Si_3N_4 (CeO_2)	1000
RS-Si_3N_4	1400
HP-SiC	1400[a]
Sintered SiC	1400[a]
Si/SiC	1200

[a] Also limited 2000 hr/1500 C exposures.

TABLE 33. POST-EXPOSURE EVALUATION PARAMETERS

Oxide Scale Morphology
- Optical
- SEM

Weight Gain

Residual Strength
- fracture origins

Scale Products
- XRD,[a] crystalline compounds
- XRF,[b] elemental

[a] X-ray diffraction.
[b] X-ray fluorescence.

TABLE 34. WEIGHT CHANGE AFTER OXIDATION EXPOSURE FOR VARIOUS HOT-PRESSED Si_3N_4 MATERIALS

Material	Exposure	Weight Change, %
Norton NCX-34 (8% Y_2O_3)	100 hr/1000°C	+0.02
	1000 hr/1000°C	+0.10
Ceradyne Ceralloy 147Y-1 (8% Y_2O_3)	100 hr/1000°C	+0.004
	1000 hr/1000°C	+0.05
Ceradyne Ceralloy 147Y (15% Y_2O_3)	100 hr/1000°C	+2.70
	1000 hr/1000°C	+0.03 to +5.1
Harbison-Walker (10% CeO_2)	100 hr/1000°C	+0.14
	1000 hr/1000°C	+0.34
Ceradyne Ceralloy 147A (1% MgO)	100 hr/1200°C	+0.17
	1000 hr/1200°C	+0.48
Fiber Materials (4% MgO)	100 hr/1200°C	+0.46
	1000 hr/1200°C	+2.62 (2.1 to 3.2)
Toshiba HP-Si_3N_4 (4% Y_2O_3, 3% Al_2O_3)	100 hr/1000°C	+0.14 (+0.1 to +0.3)
	1000 hr/1000°C	+0.06 to +0.58
Toshiba HP-Si_3N_4 (3% Y_2O_3, 4% Al_2O_3, SiO_2)	100 hr/1000°C	-0.03
	1000 hr/1000°C	-0.007
Westinghouse (4% Y_2O_3, SiO_2)	100 hr/1000°C	+0.003
	1000 hr/1000°C	+0.013

TABLE 35. WEIGHT CHANGE AFTER OXIDATION EXPOSURE FOR SINTERED Si_3N_4 MATERIALS

Material	Exposure	Weight Change, %
Rocketdyne SN-50 Sintered Si_3N_4 (6% Y_2O_3, 4% Al_2O_3)	100 hr/1000°C	+0.14
Rocketdyne SN-46 Sintered Si_3N_4 (14% Y_2O_3, 7% SiO_2)	1000 hr/1000°C	+0.64 (+4% thickness change)
AiResearch Sintered Si_3N_4 (8% Y_2O_3, 4% Al_2O_3)	100 hr/1200°C 1000 hr/1200°C	+0.08 +0.21

TABLE 36. WEIGHT CHANGE AFTER OXIDATION EXPOSURE FOR VARIOUS REACTION-SINTERED Si_3N_4 MATERIALS

Material	Exposure	Weight Change %	Unexposed Fractional Porosity		
			Open	Closed	Total
Norton 1979 NC-350	100 hr/1400°C	+1.90	0.18-0.24	.04(Avg.)	0.23-0.27
	1000 hr/1400°C	+1.57 (1.21-2.05)			
Norton 1976 NC-350	100 hr/1400°C	--	0.12	0.08	0.20
	1000 hr/1400°C	+0.13			
Ford (IM)	100 hr/1400°C	+0.63	0.11	0.03	0.14
	1000 hr/1400°C	+0.83			
AiResearch Airceram RBN-101 (Slipcast)	100 hr/1400°C	-0.25	0.06	0.04	0.10
	1000 hr/1400°C	-0.02 to +0.09			
AiResearch Airceram RBN-122 (Injection-Molded)	100 hr/1400°C	+0.74	0.11	0.06	0.17
	1000 hr/1400°C	+1.16			
Annawerk Ceranox NR-115H	100 hr/1400°C	+0.84 (0.56-1.13)	0.12-0.15	.06(Avg)	0.16-0.23
	1000 hr/1400°C	+0.98 (0.75-1.46)			
AED Nitrasil: • Batch 1	100 hr/1400°C	+2.19	0.18	0.04	0.22
	1000 hr/1400°C	+2.60			
• Batch 2	100 hr/1400°C	+1.32	0.17	0.04	0.21
	1000 hr/1400°C	+2.50			
• Batch 3	100 hr/1400°C	+1.23	0.13	0.06	0.19
	1000 hr/1400°C	+1.49			
• Batch 4	100 hr/1400°C	+1.01	0.12	0.06	0.18
	1000 hr/1400°C	+1.09			
• Batch 5	100 hr/1400°C	+0.94	0.12	0.06	0.18
	1000 hr/1400°C	+1.52			
Georgia Tech RS-Si_3N_4	100 hr/1400°C	+0.42	0.16	0.03	0.19
	1000 hr/1400°C	+1.35			
AME RS-Si_3N_4	100 hr/1400°C	+8.87 (8.32-9.37)	0.32	0.02	0.34
	1000 hr/1400°C	+8.12 (6.90-9.14)			
AiResearch RBN-104	100 hr/1400°C	+0.11	0.08	0.05	0.13
	1000 hr/1400°C	+0.18			

TABLE 37. WEIGHT CHANGE AFTER OXIDATION EXPOSURE
FOR VARIOUS SILICON CARBIDE MATERIALS

Material	Exposure	Weight Change %
Carborundum Sintered α-SiC	100 hr/1400°C	--
	1000 hr/1400°C	+0.07
General Electric Sintered SiC	100 hr/1400°C	-0.14 to +0.08
	1000 hr/1400°C	+0.01 to +0.23
	2000 hr/1500°C	0.39 to 1.21
Norton NC-203 HP-SiC (2% Al_2O_3)	100 hr/1400°C	+0.11
	1000 hr/1400°C	+0.10
	2000 hr/1500°C	+0.30
Ceradyne Ceralloy 146A HP-SiC (2% Al_2O_3)	100 hr/1400°C	+0.07
	1000 hr/1400°C	+0.20
Ceradyne Ceralloy 146I HP-SiC (2% B_4C)	100 hr/1400°C	+0.14
	1000 hr/1400°C	+0.49
UKAEA/BNF Refel Si/SiC		
• Diamond-Ground	100 hr/1200°C	+0.01
	1000 hr/1200°C	+0.08
• As-Processed	100 hr/1200°C	+0.02
	1000 hr/1200°C	+0.08
Norton NC-430 Si/SiC	100 hr/1200°C	+0.01
	1000 hr/1200°C	+0.06
Coors Si/SiC	100 hr/1200°C	+0.06
	1000 hr/1200°C	+0.16
Kyocera SC-201 Sintered SiC	100 hr/1400°C	+0.03
	1000 hr/1400°C	+0.09

TABLE 38. RETAINED STRENGTH OF HOT-PRESSED Si_3N_4 MATERIALS

Material	Non-Exposed		Static Air Exposure	Exposed	
	Mean R.T. Strength, psi	Fracture Origins		Mean R.T. Strength, psi	Fracture Origins
Norton NCX-34 (8% Y_2O_3)	126,730	Machining flaws and processing defects	100 hr/1000°C 1000 hr/1000°C	136,480 137,550	Machining flaws Undetermined
Ceradyne Ceralloy 147Y-1 (8% Y_2O_3)	83,250	Machining flaws and processing defects	100 hr/1000°C 1000 hr/1000°C	89,320 79,880	Undetermined Dark inclusion
Ceradyne Ceralloy 147Y (15% Y_2O_3)	87,850	Undetermined	100 hr/1000°C 1000 hr/1000°C	5,590 0^a 102,000	Internal cracking Undetermined
Harbison-Walker (10% CeO_2)	87,890	Primarily inclusions	100 hr/1000°C 1000 hr/1000°C	76,840 49,940	Dark inclusion Oxidation pitting
Ceradyne Ceralloy 147A (1% MgO)	87,090	Primarily inclusions	100 hr/1200°C 1000 hr/1200°C	57,590 58,250	Inclusions and oxidation pits Oxidation pits
Fiber Materials (4% MgO)	73,680	Dark-colored inclusions	100 hr/1200°C 1000 hr/1200°C	66,110 42,980	Oxidation pits Oxidation pits
Toshiba HP-Si_3N_4 (4% Y_2O_3, 3% Al_2O_3)	105,140	Dark, shiny inclusions	100 hr/1000°C 1000 hr/1000°C	102,410 112,730	Dark, shiny inclusions Dark, shiny inclusions
Toshiba HP-Si_3N_4 (3% Y_2O_3, 4% Al_2O_3, SiO_2)	84,100	Inclusions	100 hr/1000°C 1000 hr/1000°C	65,270 42,710	Inclusions Inclusions } No breaks at joints
Westinghouse (4% Y_2O_3, SiO_2)	90,980	Dark. shiny inclusions	100 hr/1000°C 1000 hr/1000°C	112,050 93,730	Grain or inclusion Machining defect

Note: The typical population of oxidized test bars for retained 4-point bend strength measurement was two to four samples.

^{a}0 psi: Gross macrocracking due to accelerated oxidation. Sample failed in furnace due to destructive phase instability.

TABLE 39. RETAINED STRENGTH OF SINTERED Si_3N_4 MATERIALS

Material	Non-Exposed		Static Air Exposure	Exposed	
	Mean R.T. Strength, psi	Fracture Origins		Mean R.T. Strength, psi	Fracture Origins
Rocketdyne SN-50 Sintered Si_3N_4 (6% Y_2O_3, 4% Al_2O_3)	51,600	Porosity open to tensile surface	100 hr/1000°C	45,270	Inclusions and accelerated oxidation at surface pores
Rocketdyne SN-46 Sintered Si_3N_4 (14% Y_2O_3, 7% SiO_2)	51,600	Porosity open to tensile surface	1000 hr/1000°C	34,450	Oxidation at pre-existing surface pores
AiResearch Sintered Si_3N_4 (8% Y_2O_3, 4% Al_2O_3)	79,450	Surface and Subsurface porosity	100 hr/1200°C	61,900	Surface porosity
			1000 hr/1200°C	59,190	Surface porosity

Note: The typical population of oxidized test bars for retained 4-point bend strength measurement was two to four samples.

TABLE 40. RETAINED STRENGTH OF REACTION SINTERED Si_3N_4 MATERIALS

Material	Non-Exposed		Static Air Exposure	Exposed	
	Mean R T Strength, psi	Fracture Origins		Mean R T Strength, psi	Fracture Origins
Norton 1979 NC-350	37,560	Primarily subsurface inclusions	100 hr/1400°C 1000 hr/1400°C	37,690 34,480	Subsurface inclusions; Primarily subsurface inclusions
Norton 1976 NC-350	29,450	Pores, pore agglomerates	100 hr/1400°C 1000 hr/1400°C	36,600 30,360	Primarily oxidation pits Primarily oxidation pits, some undeter
Ford (IM)	38,180	Processing defects, inclusions and porosity	100 hr/1400°C 1000 hr/1400°C	33,060 25,280	Oxidation pits Primarily oxidation pits
AiResearch Airceram RBN-101 (Slipcast)	37,920	Primarily inclusions	100 hr/1400°C 1000 hr/1400°C	22,990 30,590	Oxidation pits Oxidation pits
AiResearch Airceram RBN-122 (Injection-Molded)	32,580	Inclusions, some with adjoining pores	100 hr/1400°C 1000 hr/1400°C	18,390 5,960	Oxidation pits Primarily oxidation pits
Annawerk Ceranox NR-115H	28,870	Inclusions	100 hr/1400°C 1000 hr/1400°C	21,890 0,[a]19,100	Primarily oxidation pits Primarily oxidation pits
AED Nitrasil: • Batch 1	29,920	Inclusions, some with adjoining pores	100 hr/1400°C 1000 hr/1400°C	20,680 15,980	Oxidation pits Undetermined
• Batch 2	27,400	Inclusions, some with adjoining pores	100 hr/1400°C 1000 hr/1400°C	19,060 16,710	Oxidation pits and inclusions Oxidation pits or undetermined
• Batch 3	28,190	Inclusions or subsurface pores	100 hr/1400°C 1000 hr/1400°C	17,800 11,840	Oxidation pits Undetermined
• Batch 4	30,730	Inclusions, some with adjoining pores	100 hr/1400°C 1000 hr/1400°C	18,690 14,730	Primarily oxidation pits Primarily oxidation pits
• Batch 5	26,080	Primarily dark inclusions	100 hr/1400°C 1000 hr/1400°C	18,990 21,210	Surface and subsurface porosity Pore under oxide skin
Georgia Tech RS-Si_3N_4	26,220	Primarily light-colored inclusions	100 hr/1400°C 1000 hr/1400°C	26,240 19,560	Oxidation pit Oxidation pit
AME RS-Si_3N_4	7,470	Undetermined due to porous nature of material	100 hr/1400°C 1000 hr/1490°C	1,400 1,750	Undetermined Undetermined
AiResearch RBN-104	40,270	Subsurface inclusions and pores	100 hr/1400°C 1000 hr/1400°C	33,640 36,590	Primarily oxidation pits Primarily oxidation pits

Note: The typical population of oxidized test bars for retained 4-point bend strength measurement was two to four samples.

[a] 0 psi: Gross macrocracking due to destructive phase instability. Sample failed in furnace under zero stress.

TABLE 41. RETAINED STRENGTH OF SiC MATERIALS

Material	Non-Exposed		Exposed		
	Mean R.T. Strength, psi	Fracture Origins	Static Air Exposure	Mean R.T. Strength, psi	Fracture Origins
Carborundum Sintered α-SiC	44,230	Primarily inclusions or large grains	100 hr/1400°C 1000 hr/1400°C	50,160 42,500	Subsurface inclusion or large grain. Oxidation pits.
General Electric Sintered SiC	63,770	Primarily inclusions or large grains	100 hr/1400°C 1000 hr/1400°C 2000 hr/1500°C	73,650 59,200 49,970	Undetermined. Internal processing defects. Oxidation pits.
Norton NC-203 HP-SiC (2% Al$_2$O$_3$)	101,810	Undetermined	100 hr/1400°C 1000 hr/1400°C 2000 hr/1500°C	91,890 91,470 103,670	Chips at non-chamfered edge Undetermined. Undetermined.
Ceradyne Ceralloy 146A HP-SiC (2% Al$_2$O$_3$)	60,960	Undetermined	100 hr/1400°C 1000 hr/1400°C	43,880 54,010	Undetermined. Undetermined.
Ceradyne Ceralloy 146I HP-SiC (2% B$_4$C)	45,620	Some machining damage, primarily undetermined	100 hr/1400°C 1000 hr/1400°C	43,910 43,610	Undetermined. Oxidation pits
UKAEA/BNF Refel Si/SiC					
• Diamond-Ground	44,920	Large grains	100 hr/1200°C 1000 hr/1200°C	a 63,890	Mixed internal processing defects and undetermined. Surface oxide pits and large grains.
• As-Processed	33,600	Some large grains, primarily undetermined	100 hr/1200°C 1000 hr/1200°C	43,560 38,770	Undetermined. Undetermined.
Norton NC-430 Si/SiC	30,420	Undetermined	100 hr/1200°C 1000 hr/1200°C	29,640 40,110	Undetermined. Undetermined.
Coors Si/SiC	50,620	Mostly undetermined, some machining damage	100 hr/1200°C 1000 hr/1200°C	44,800 27,120	Undetermined. Oxidation
Kyocera SC-201 Sintered SiC	56,080	Primarily undetermined, some surface and subsurface pores	100 hr/1400°C 1000 hr/1400°C	59,540 50,720	Undetermined Oxidation pits (surface pores?)

Note: The typical population of oxidized test bars for retained 4-point bend strength measurement was two to four samples.

[a] Not all samples were standard (1/4 x 1/8 x 2-1/4 in.) dimensions. Dimensions and strengths were:

inches	psi
1/4 x 1/8 x 2	63,960
1/4 x 1/4 x 2	50,590
1/8 x 9/32 x 2-1/2	42,330
1/8 x 1/4 x 2-1/4	78,790

TABLE 42. SUMMARY OF WEIGHT CHANGE DATA

Material	Weight Gain, %
Si_3N_4 (Exposure: 1000 hr/1200°C)	
HP-Si_3N_4 (1% MgO)	0.48
HP-Si_3N_4 (4% MgO)	2.6
Sintered Si_3N_4 (8% Y_2O_3, 4% Al_2O_3)	0.21
HP-Si_3N_4 (Exposure: 1000 hr/1000°C	
HP-Si_3N_4 (8% Y_2O_3)	0.05-0.10
HP-Si_3N_4 (15% Y_2O_3)	0.03-5.1
HP-Si_3N_4 (4% Y_2O_3, 3% Al_2O_3)	0.06-0.58
HP-Si_3N_4 (3% Y_2O_3, 4% Al_2O_3, SiO_2)	0
HP-Si_3N_4 (4% Y_2O_3, SiO_2)	0.01
HP-Si_3N_4 (10% CeO_2)	0.34
SiC Material (Exposure: 1000 hr/1400°C	
HP-SiC (2% A_2O_3)	0.1-0.2
HP-SiC (2% B_4C)	0.49
Sintered SiC	0.08 (α)
	0.01-0.23 (β)
Si/SiC[a]	0.06-0.16

[a]1200°C exposure.

TABLE 43. SUMMARY OF STRENGTH CHANGES DUE TO 1000 HR STATIC AIR EXPOSURE

Material	Exposure Temp., °C	Strength Change/25°C	Fracture Origins Unexposed	Fracture Origins Exposed
A. HP-Si$_3$N$_4$ Materials				
HP-Si$_3$N$_4$ (1% MgO)	1200	-33%	Inclusions	Oxidation pits
HP-Si$_3$N$_4$ (4% MgO)	1200	-41%	Inclusions	Oxidation pits
HP-Si$_3$N$_4$ (8% Y$_2$O$_3$)	1000	+7%	Inclusions	Inclusions
HP-Si$_3$N$_4$ (15% Y$_2$O$_3$)	1000	+10% -100%	Inclusions	Internal cracking due to accelerated oxidation
HP-Si$_3$N$_4$ (4% Y$_2$O$_3$, 3% Al$_2$O$_3$)	1000	No change	Inclusions	Inclusions
HP-Si$_3$N$_4$ (4% Y$_2$O$_3$, SiO$_2$)	1000	+10%	Inclusions	Inclusions
HP-Si$_3$N$_4$ (10% CeO$_2$)	1000	-50%	Inclusions	Oxidation pits
B. RS-Si$_3$N$_4$ Materials				
Reaction sintered Si$_3$N$_4$	1400	+3% to -60%	Inclusions, porosity	Primarily oxidation pits
C. SiC Materials				
HP-SiC (2% Al$_2$O$_3$)	1400 1500[a]	-10% 0	Undetermined	Undetermined
HP-SiC (2% B$_4$C)	1400	-4%	Undetermined	Oxidation pits
Sintered α-SiC	1400	-8%	Inclusions, grains, pores	Oxidation pits
Si/SiC	1200	+30% to -46%	Undetermined	Mostly undetermined, some oxidation pits

[a] 2000 hr.

the oxidation of silicon-base ceramics found in the literature.[69-91] The overview summary for each material type is presented below.

11.1 Si_3N_4 MATERIALS

The hot-pressed silicon nitride materials studied on this program have been doped with MgO, Y_2O_3, CeO_2, Y_2O_3-Al_2O_3, Y_2O_3-SiO_2, and Y_2O_3-Al_2O_3-SiO_2.

11.1.1 MgO-Doped HP-Si_3N_4

MgO-doped HPSN materials exhibit more rapid oxidation than Y_2O_3-doped materials. This is directly observed in the weight gain data for 1000 hr/1200°C exposure shown in Table 42. The Ceradyne and FMI materials contained 1% and 4% MgO additive, respectively. Table 42 illustrates the decreased resistance to oxidation as the additive content increases (i.e., increased weight gain). The surface scale morphology of these materials is presented in Figures 76 through 79. The SEM detects the thick crystalline scale products after 100 hr exposure. After 1000 hr exposure the scales are easily observed in the low magnification optical microscope.

Such MgO-doped materials form a SiO_2 scale upon oxidation which is quickly modified by the impurities in the unexposed material. These cation impurities are tabulated in the spectrographic analysis results presented in Table 7. Typically Mg, Al, Fe, Ca, and Mn impurities diffuse up through the grain boundary into the oxide scale where they either form mixed crystalline silicates or dissolve into the glass phase. This outward diffusion of impurity (Mg and Ca) cations from the amorphous grain boundary phase through the surface oxide film is thought to be the rate-controlling process in the oxidation of MgO-doped Si_3N_4.[69] The main crystalline scale product is usually $MgSiO_3$ (enstatite), with small amounts of cristobalite. These expected results were confirmed for the FMI and Ceradyne MgO-doped HPSN materials. X-ray diffraction (XRD) analysis for both these materials indicated the major scale products to be $MgSiO_3$ (enstatite,

Materials Evaluation 213

200X

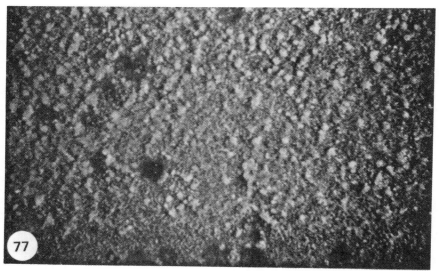

12X

Figure 76. Surface of Ceradyne Ceralloy 147A, HP-Si$_3$N$_4$ (1% MgO) sample after 100 hr/1200°C static air exposure.

Figure 77. Surface of Ceradyne Ceralloy 147A HP-Si$_3$N$_4$ (1% MgO) sample after 1000 hr/1200°C static air exposure.

214 *Ceramic Materials for Advanced Heat Engines*

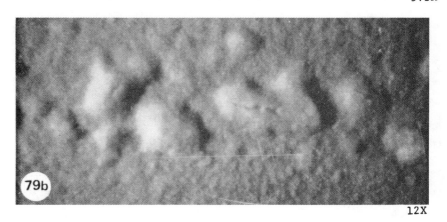

Figure 78. Surface of Fiber Materials, Inc., HP-Si_3N_4 (4% MgO) sample after 100 hr/ 1200°C static air exposure.

Figure 79. Surface of Fiber Materials, Inc., HP-Si_3N_4 (4% MgO) sample after 1000 hr/1200°C static air exposure.

which is orthorhombic, and clinoenstatite, which is monoclinic), and the minor crystalline scale product to be cristobalite. The X-ray fluorescence (XRF) scan of the oxide scale is shown in Figure 80. When these scale impurities are compared to the spectrographic analysis data (Table 7), we find that most of the major matrix material impurities have diffused up into the oxide scale. An exception is the original Al impurity that has apparently not migrated into the oxide scale. Phosphorus was not a major impurity detected by spectrographic analysis, and therefore its presemce in the XRF scan of the oxide scale is not understood at this time.

At short oxidation times, distinct oxide mounds are formed on the Si_3N_4 surface as shown in Figure 81a. The mound formation often contains a hole in the top which presumably was formed by the evolution of N_2 gas from the SiO_2/Si_3N_4 interface. At longer times $MgSiO_3$ is formed as discussed above. It is believed that $MgSiO_3$ reacts with Si_3N_4 to form pits at the scale/matrix interface. These pits become fracture origins as illustrated in Figure 81b. Table 38 illustrates for both Ceradyne 1% MgO- and FMI 4% MgO-modified Si_3N_4 materials, that oxidation pits were the fracture origins for 100 and 1000 hr exposure at 1200°C. Such oxidation can be responsible for as much as 40% strength loss.

11.1.2 CeO_2-Doped HP-Si_3N_4

CeO_2 dopants for HP-Si_3N_4 are being investigated in an attempt to form more deformation resistant intergranular phases than the amorphous grain boundary films present in MgO-doped materials. Harbison-Walker HP-Si_3N_4 (10% Ce_2O) was exposed in static air for 1000 hr at 1000°C. The weight gain data given in Table 42 indicate much more rapid oxidation for Ce_2O-containing silicon nitride compared to Y_2O_3-modified material. No direct comparison can be made with MgO-doped materials since the exposure temperatures were different. The strength and fracture origin data of Table 38 also illustrate the poor oxidation resistance when compared to Y_2O_3-containing materials. While good strength retention was obtained for the Harbison-Walker material

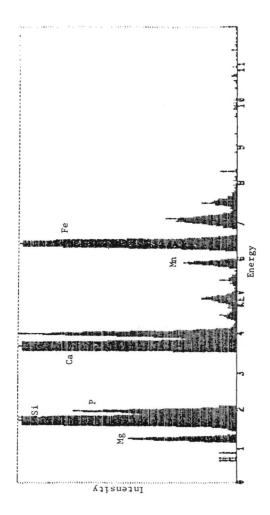

Figure 80. X-ray fluorescence scan of oxidation scale for Ceradyne Ceralloy 147A HP-Si_3N_4 (1% MgO) after 1000 hr/1200°C static air exposure.

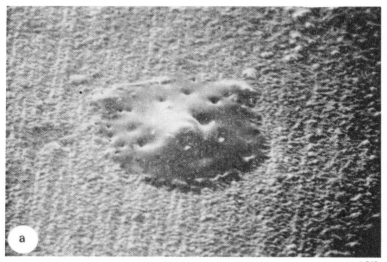

200X

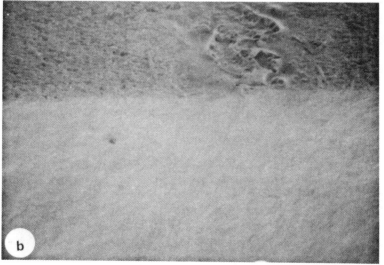

92X

Figure 81. Oxidation of MgO-doped HP-Si_3N_4.

(a) Oxide mound on tensile surface of Norton NC-132 (1% MgO) after 15 min/1350°C exposure. (Photo courtesy of R. W. Rice, NRL.)

(b) Oxidation pit as fracture origin in Ceradyne Ceralloy 147A (1% MgO) exposed for 100 hr/1200°C

after 100 hr exposure (Table 38), excessive oxidation pitting caused a 50% strength loss after 1000 hr exposure at 1000°C. The morphology of the oxide scale after 100- and 1000-hr exposures at 1000°C is shown in Figures 82 and 83. The crystalline scales shown were found by X-ray diffraction to be mainly cristobalite. No crystalline cerium compounds were identified. Figure 84 presents the XRF elemental scan of the oxide scale. The Ce present is apparently dissolved in an amorphous phase on the surface. The iron and calcium impurities present in the scale were in the unexposed material according to spectrographic analysis (Table 7). These impurities in ceria-containing Si_3N_4 diffuse to the oxide scale as they do in magnesia-doped material. As with MgO-modified Si_3N_4, the aluminum impurity is not detected in the surface oxide scales of CeO_2-doped material, as shown in Figure 84. Lange[81] discusses that ceria-containing silicon nitride may have unstable crystalline phases that experience accelerated oxidation in a manner similar to the Y_2O_3-doped materials discussed below. No such crystalline cerium phases were detected by X-ray diffraction for the Harbison-Walker material (Table 6). However, their existence might help to explain the relatively poor oxidation resistance exhibited by this material at 1000°C.*

11.1.3 Y_2O_3-Doped HP-Si_3N_4

Y_2O_3 has proven to be a promising densification aid for hot-pressed silicon nitride. Its main advantage is that an intergranular phase is formed that can be crystallized, thus improving high-temperature strength and creep resistance when compared to MgO-doped Si_3N_4 materials, which have an amorphous grain boundary phase. We have demonstrated above that much less subcritical crack growth is evident in many Y_2O_3-doped Si_3N_4 materials.

---

*It is noted, however, that in-house work at AFWAL indicates that much better oxidation resistance is obtained with lower CeO_2 content Si_3N_4 materials.

Materials Evaluation 219

SEM, 1000X

SEM, 1000X

Figure 82. Surface of Harbison-Walker HP-Si_3N_4 (10% CeO_2) sample after 100 hr/1000°C static air exposure.

Figure 83. Surface of Harbison-Walker HP-Si_3N_4 (10% CeO_2) sample after 1000 hr/1000°C static air exposure.

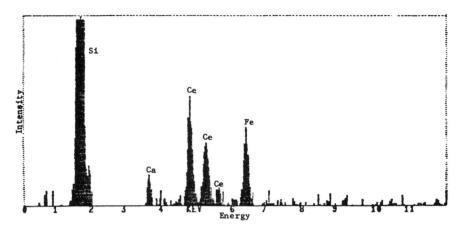

Figure 84. X-ray fluorescence scan of oxidation scale for Harbison-Walker HP-Si_3N_4 (10% CeO_2) after 1000 hr/1000°C static air exposure.

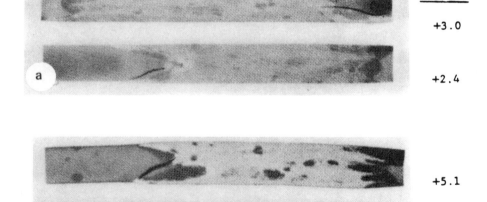

Figure 85. Photographs of Ceradyne HP-Si_3N_4 (15% Y_2O_3) samples after static air exposure.

(a) 100 hr/1000°C
(b) 1000 hr/1000°C

We have also demonstrated that such Y_2O_3-containing materials can have interesting oxidation behavior. Catastrophic results such as illustrated in Figure 85 for a 15% Y_2O_3-doped Si_3N_4 can result at temperatures between ~750° and 1000°C. It is thought that such behavior is caused by accelerated linear oxidation of $Si_3Y_2O_3N_4$, $YSiO_2N$, or $Y_{10}Si_7O_{23}N_4$ phases in these materials. Such materials exist in the Si_3N_4-SiO_2-Y_2O_3 phase fields labeled A, B, and C in Figure 86.[69] However, if the processed material is in the phase triangle containing crystalline Si_2N_2O and $Y_2Si_2O_7$ phases (i.e., phase field D in Figure 86), then parabolic kinetics are observed and the oxidation rates are very low. Other mechanistic interpretations of this effect involve the influence of W and WC contamination[86] and carbon impurity effects.[85].

Several Y_2O_3-containing materials have been investigated on this program, and the results are consistent with this mechanistic interpretation. The Ceradyne 15% Y_2O_3-doped material that exhibited the catastrophic oxidation (shown in Figure 85) indeed contained $YSiO_2N$ as a minor crystalline phase, as shown in the X-ray diffraction results presented in Table 6. The Westinghouse material, however, doped with 4% Y_2O_3 and an unknown amount of SiO_2, contained $Y_2Si_2O_7$ as a minor phase, and did not develop nonprotective oxide scales when exposed at 1000°C. The appearance of the Westinghouse material after 100 and 1000 hr exposure at 1000°C is shown in Figure 87. Figure 88 presents an SEM view of the surface of Norton NCX-34 (8% Y_2O_3-doped) after 1000 hr/ 1000°C. This material only exhibited a slight evidence of the phase instability, better seen in the optical photograph in Figure 89. Thus, the goal in processing these materials is to keep the Y_2O_3 content lower than ~6-8%. Several attempts are currently being made with a 4% Y_2O_3 content, similar to the Westinghouse material.

The weight change and strength/fracture origin information presented in Tables 34-37 and Tables 38-41, respectively, illustrates that if the catastrophic oxidation effect is avoided, then

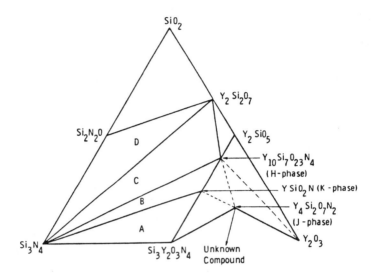

Figure 86. Phase relationships in the Si_3N_4-SiO_2-Y_2O_3 system at 1600°-1750°C (after Singhal).[69]

Materials Evaluation 223

Figure 87. Surfaces of Westinghouse HP-Si_3N_4 (4% Y_2O_3, SiO_2) after 1000°C static air exposure.
(a) 100 hr/1000°C exposure
(b) 1000 hr/1000°C exposure

224 Ceramic Materials for Advanced Heat Engines

1000X

12X

Figure 88. SEM view of surface of Norton NCX-34 HP-Si$_3$N$_4$ (8% Y$_2$O$_3$) sample after 1000 hr/1000°C static air exposure.

Figure 89. Optical macrograph of surface of Norton NCX-34 HP-Si$_3$N$_4$ (8% Y$_2$O$_3$) sample after 1000 hr/ 1000°C static air exposure.

Y_2O_3-containing materials have good oxidation resistance. Weight gain during oxidation is very low, and the exposed strength is equivalent to the unexposed strength. This is mainly evident for the 8% Y_2O_3-doped materials from Norton and Ceradyne. It is further noted in Table 38 that exposed and non-exposed fracture origins were the same for these two materials, which is indicative of the absence of oxidation pitting. The oxide scales on these materials were too thin to remove for XRD and XRF analysis of the scale products.

Toshiba has developed a very significant HP-Si_3N_4 doped with 4% Y_2O_3 and 3% Al_2O_3. The high strength of this matarial was discussed above. The weight gain data in Table 34 show that the addition of Al_2O_3 has decreased the oxidation resistance when compared to materials modified by Y_2O_3 only, as evidenced by larger weight gain. However, when SiO_2 is also added, with or without Al_2O_3, the oxidation resistance is very good. This is illustrated by the weight gain data in Table 34 for the Toshiba (3% Y_2O_3, 4% Al_2O_3, SiO_2) modified Si_3N_4 and the Westinghouse 4% Y_2O_3-SiO_2 modified Si_3N_4. In general, the nonexposed bend strengths were maintained for these materials, and fracture origins remained unchanged. Figures 90 and 91 illustrate the nature of the oxide scales on the two Y_2O_3-Al_2O_3-doped Toshiba materials, with and without SiO_2 addition. Oxide scale products could only be identified for the Toshiba HP-Si_3N_4 (4% Y_2O_3, 3% Al_2O_3) material. X-ray diffraction indicated that the scale consisted of cristobalite. No crystalline aluminum silicate or mullite compounds were identified.

11.1.4 RS–Si_3N_4 Materials

The properties of RS-Si_3N_4 materials can vary widely, and depend on the specific nature of the porosity, pore size, and pore size distribution. Impurities only have an effect in some cases, since RSSNs are generally very pure, containing typically 0.5-1% cation impurities (spectrographic analysis, Table 7). As discussed by Singhal,[70] oxidation in RS-Si_3N_4 materials occurs in two stages. First, internal oxidation occurs within the network

12X

12X

Figure 90. Surfaces of Toshiba HP-Si$_3$N$_4$ (4% Y$_2$O$_3$, 3% Al$_2$O$_3$) samples after 1000°C static air exposure.

(a) 100 hr/1000°C exposure
(b) 1000 hr/1000°C exposure

12X

↙――――Joint

12X

Figure 91. Surfaces of Toshiba HP-Si$_3$N$_4$ (3% Y$_2$O$_3$, 4% Al$_2$O$_3$, SiO$_2$) sample after 1000°C static air exposure.
(a)   100 hr/1000°C exposure
(b)   1000 hr/1000°C exposure

of open porosity until the pores are completely filled, or until the surface is sealed, which limits the supply of oxygen to the interior. The second stage of oxidation then occurs only on the external surface. In both stages, oxidation follows parabolic behavior due to the diffusion-controlled process of inward oxygen transport through the SiO_2 scale to the Si_3N_4/SiO_2 interface. The oxidation rates are higher in the first stage, since the surface area is much greater (the interconnected open porosity is exposed). For many $RS-Si_3N_4$ materials, experience has shown that up to 1100°C most of the oxidation is internal. However, at ~1200°C a transition occurs from first to second stage oxidation at very short times. However, much internal oxidation can occur prior to the transition at temperatures of 1200°C and above. We have demonstrated this for the very porous AME material (ρ = 2.1 g cm^{-3}). After 15 min/1400°C exposure, the oxide penetrated half the thickness of the test sample. After 100 hr/1400°C exposure the oxide had penetrated through all of the open porosity. This may be an extreme example, since this material contained very large open porosity, i.e., 32%. After 1000 hr exposure, significant second-stage external oxidation had occurred in this material, since cracks were observed in the scale upon cooling, as shown in Figure 92.

The higher density (ρ = 2.5-2.8 g cm^{-3}) $RS-Si_3N_4$ materials studied on this program were also exposed for 100 and 1000 hr at 1400°C. The transition to second stage oxidation apparently occurred rapidly, since only surface scales were observed, with no major internal attack. Oxidation was still controlled by the open porosity though, in the sense that higher open porosity means a greater Si_3N_4 surface area at the Si_3N_4/SiO_2 interface. This is illustrated in Figure 93, where percent weight gain is plotted as a function of fractional open porosity for all RSSN materials oxidized for 1000 hr at 1400°C. The tabular data are presented in Table 36. No such correlation was found when weight gain data were plotted as a function of total metallic impurity content, or the amount of any specific impurity present (e.g., Al

Materials Evaluation 229

SEM, 200X

Figure 92. Surface of AME RS-Si$_3$N$_4$ sample after 1000 hr/1400°C static air exposure.

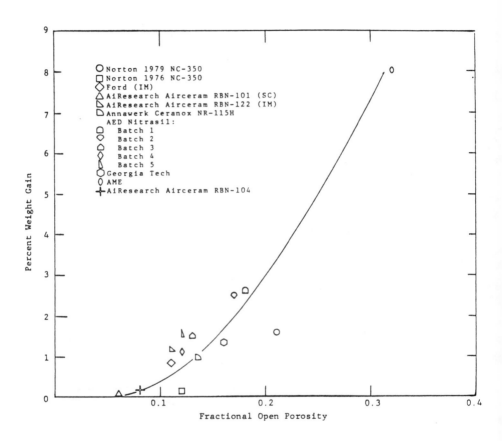

Figure 93. Weight gain-porosity relation for RS-Si$_3$N$_4$ materials exposed in static air for 1000 hr at 1400°C.[4]

or Fe). Thus, oxidation in RS-Si_3N_4 materials, as measured by weight gain, is determined by the open porosity.

The residual strength of oxidized RS-Si_3N_4 materials has been found to be either lower or higher than the unexposed strength, depending on the extent of internal and external oxidation, the nature of the SiO_2 formed, the temperature of the strength measurement, and whether or not any thermal cycling had occurred. In general, if only a small amount of internal oxidation occurs, we would expect a strength increase due to rounding off of internal pores. If gross internal oxidation occurs, we would expect to see internal cracking upon cooling as was seen with the AME material. If stage two oxidation occurred (i.e., if a thick cristobalite layer is formed on the surface), then we would expect the high-temperature strength to increase due to blunting of surface pore features, and the room-temperature strength to decrease due to cracking of the cristobalite layer upon cooling through its 250°C phase inversion.

The residual strength data for all RS-Si_3N_4 materials exposed for 100 and 1000 hr at 1400°C are contained in Table 40 and summarized in Table 43. Most materials experienced sufficient oxidation scale cracking and pitting to result in up to 60% reduction in strength when compared to unexposed material. An SEM view of a typical surface scale after 1000 hr/1400°C static air exposure is shown in Figure 94. The two Norton NC-350 materials (1976- and 1979-vintage) are particularly interesting materials. The 1976 NC-350 experienced no reduction in strength after 1000 hr/1400°C exposure. However, fracture origins were oxidation pits, and the surface scale was cracked as shown in Figure 95. The 1979 NC-350 experienced only marginal strength loss after similar exposure as shown in Table 40. However, Table 36 illustrates that the 1979 material had a much greater open porosity, and experienced an order of magnitude greater weight gain for both exposure times. It is notable that 1979 NC-350 experienced no change in fracture origins due to exposure. As shown in Table 40, subsurface inclusions were the fracture origins under all

100X

SEM, 500X

Figure 94. Surfaces of AiResearch injection molded RS-Si$_3$N$_4$ (Airceram RBN-122) samples after 1000 hr/1400°C static air exposure.

Figure 95. Surface of 1976 vintage Norton NC-350 RS-Si_3N_4 after 1000 hr/1400°C static air exposure.

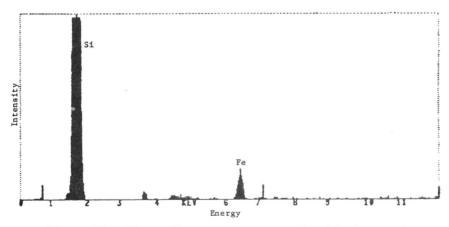

Figure 96. X-ray fluorescence scan of oxidation scale for **Norton NC-350** RS-Si_3N_4 after 1000 hr/1400°C static air exposure.

test conditions for this material. Thus, even though 1979 NC-350 experienced a relatively large weight gain upon exposure, the strength and critical flaw structure remained unchanged.

The crystalline oxide scale was identified for all RS-Si_3N_4 materials by XRD to be cristobalite. Elemental X-ray fluorescence scans indicated Fe was typically present in the scales. An XRF scan for NC-350 scale is shown in Figure 96. The spectrographic analysis results in Table 7 indicate that iron is the major impurity in RS-Si_3N_4. Figure 97 illustrates the surface scale of AED Nitrasil after 1000 hr/1400°C exposure. These two batches of material had a total of almost 2% Fe, Al, and Ca impurities (Table 7). The surface scale for a newer batch of Nitrasil is shown in Figure 98. Figure 99 illustrates an anomalous effect for Annawerk Ceranox material that was discussed previously. The sample shown in Figure 99 spontaneously fractured during 1000 hr/1400°C static air exposure. No definite explanation for this is currently available. However, an inclusion-Si_3N_4 thermal expansion mismatch seems a plausible cause.

11.2 SiC MATERIALS

Silicon carbide materials have much better oxidation resistance than Si_3N_4 materials. Two views are found in the literature regarding the rate-controlling mechanisms of oxidation in SiC. During oxidation CO gas is formed at the SiC-SiO_2 interface, and its desorption from the interface has been found by some investigators to control the rate of oxidation.[69] Others contend that oxidation kinetics are determined by the rate of inward oxygen diffusion through the surface SiO_2 layer.[77] All investigations have found that processing additives such as Al_2O_3, B, and B_4C increase oxidation rates over additive-free CVD-SiC, but that parabolic kinetics are still observed. The additives are thought to decrease the viscosity of the surface SiO_2 layer, which results in increased oxygen diffusion through the scale, leading to greater rates of oxidation. The oxide scales are typically cristobalite, and an amorphous phase in

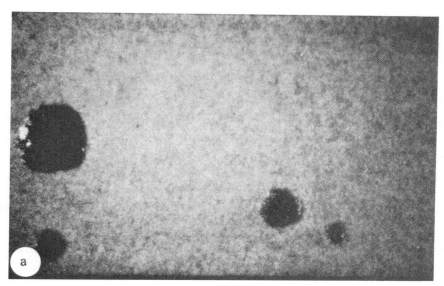

12X

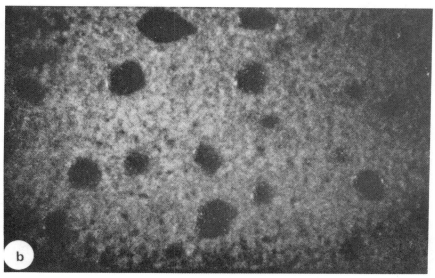

12X

Figure 97. Surface of AED Nitrasil RS-Si$_3$N$_4$ after 1000 hr/1400°C static air exposure.
(a) Batch 1
(b) Batch 2

Figure 98. Surfaces of AED Nitrasil (batch 5) RS-Si_3N_4 samples after 1400°C static air exposure.
(a) 100 hr/1400°C exposure
(b) 1000 hr/1400°C exposure

Materials Evaluation 237

5.3X

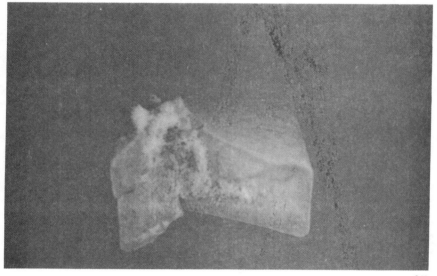
8X

Figure 99. Spontaneous fracture of Annawerk
Ceranox NR-115H RS-Si_3N_4 sample during
1000 hr/1400°C static air exposure.

which impurities such as Al and Fe concentrate. Boron impurities may be found in the scale also. Thus the outward diffusion of impurities from the matrix into the scale is analogous to the case for hot-pressed silicon nitride. In Al_2O_3-doped HP-SiC some mullite (3 $Al_2O_3 \cdot 2$ SiO_2) and aluminum silicate (Al_2SiO_5) are sometimes found in the surface scale, primarily for oxidation temperatures below 1300°C.[69]

Various hot-pressed, sintered, and silicon-infiltrated forms of silicon carbide have been exposed in static air environment on this program. The usual exposure was for 100 and 1000 hr at 1200° or 1400°C. Selected materials were exposed for 2000 hr at 1500°C. The weight change data were presented in Table 37 and summarized in Table 42.

11.2.1 HP-SiC

Hot-pressed and sintered materials were exposed at 1400°C. The use of B_4C instead of Al_2O_3 in HP-SiC is seen to result in up to a factor of five increase in weight gain. The optical photographs in Figure 100 and 101 illustrate a much thicker and more crystalline surface scale for the B_4C-doped material, even though the exposure conditions for that material were not as severe as for the Al_2O_3-doped material (1000 hr/1400°C vs. 2000 hr/1500°C). The surface scale for the B_4C-doped SiC was identified as cristobalite. Calcium and iron were found in the scale by X-ray fluorescence. The element boron cannot be detected with the equipment used. The surface scales on 2% Al_2O_3-doped HP-SiC were very thin, smooth, and shiny. The shiny nature of the scale in these materials has been thought to be indirect evidence that the scale was liquid at the exposure temperature. This coincides with the concept of Al matrix impurities entering the SiO_2 scale and reducing its viscosity. SEM views of the surface scales of such HP-SiC materials containing 2% Al_2O_3 are presented in Figures 102 and 103. The hole in the scale, shown in Figure 102, was presumably made by escaping CO gas. Note the cracking and spallation of the scale shown in Figure 103. Even though the B_4C-doped HP-SiC developed a thicker, more crystalline scale than did the

12X

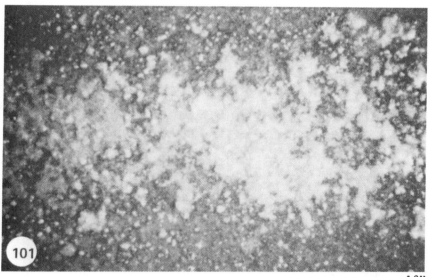

12X

Figure 100. Optical view of surface of Norton NC-203 HP-SiC (2% Al_2O_3) after 2000 hr/1500°C static air exposure.

Figure 101. Optical view of surface of Ceradyne Ceralloy 146I HP-SiC (2% B_4C) after 1000 hr/1400°C static air exposure.

240   *Ceramic Materials for Advanced Heat Engines*

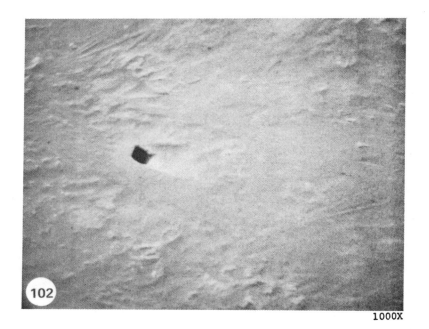

Figure 102. SEM view of surface of Norton NC-203 HP-SiC (2% Al_2O_3) after 2000 hr/ 1500°C static air exposure.

Figure 103. SEM view of surface of Ceradyne Ceralloy 146A HP-SiC (2% Al_2O_3) after 1000 hr/1400°C static air exposure.

Al_2O_3-doped material, Table 41 shows that its bend strength was not as affected by the exposure. For 1000 hr/1400°C exposure, the Al_2O_3-doped material retained ~90% of its unexposed strength, whereas the B_4C-doped material retained about 95% of its original strength. It is interesting to note that even after the 2000 hr/1500°C exposure, Norton NC-203 has the same room-temperature strength as in nonexposed material.

11.2.2 Sintered SiC

Three sintered SiC materials were evaluated: Carborundum (1979) α-SiC, Kyocera α-SiC, and General Electric β-SiC. They were exposed for 100 and 1000 hr at 1400°C. Table 42 summarized the weight change data. The α-form generally has about a factor of two lower weight gain than the best hot-pressed material (i.e., Al_2O_3 doped). This illustrates the adverse effect of impurities on oxidation resistance. The sintered materials are very pure, typically containing only ~0.5% metallic impurity, which is mainly the boron that is used as a sintering aid. The sintered β-SiC was a bit more variable, as shown in Table 42. Some samples experienced much more crystalline scale formation than the α-SiC materials, which were very uniform.

The oxide surface scales on sintered α-SiC are very thin and dull, as shown in the optical photographs in Figures 104 and 105. The SEM view of the scale formation on the Carborundum and Kyocera materials provided in Figures 106 and 107 are much more informative. The holes in the scale of the Carborundum material were presumably made by escaping CO gas. The surface of the Kyocera material appears similar. After 100 hr oxidation at 1400°C, both materials gain ~5-10% in strength, presumably due to blunting of machining marks or isolated surface porosity. After 1000 hr exposure, both α-SiC materials lose about 5-10% of their original unexposed strength, oxidation pits becoming the predominant fracture origins. The β-SiC behaves in a similar manner.

242   *Ceramic Materials for Advanced Heat Engines*

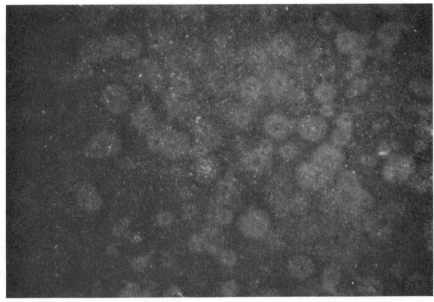

12X

Figure 104. Optical view of surface of Carborundum sintered α-SiC after 1000 hr/1400°C static air exposure.

Materials Evaluation 243

12X

12X

Figure 105. Optical view of surfaces of Kyocera SC-201 sintered α-SiC after 1400°C static air exposure.

(a) 100 hr/1400°C exposure
(b) 1000 hr/1400°C exposure

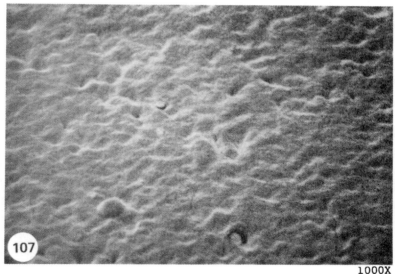

Figure 106. SEM view of surface of Carborundum sintered α-SiC after 1000 hr/1400°C static air exposure.

Figure 107. SEM view of surface of Kyocera SC-201 sintered α-SiC after 1000 hr/1400°C static air exposure.

11.2.3 Silicon-Infiltrated SiC

Silicon-infiltrated (or siliconized) SiC materials contain typically 5-15% free Si metal to densify the originally porous structure. They contain nominally 0.5% other metals (impurities), mainly aluminum and iron. Refel, Norton NC-430, and Coors 1979 SC-1 Si/SiC were exposed for 100 and 1000 hr at 1200°C. Table 42 illustrates that they experienced about the same weight gain as HP-SiC (2% Al_2O_3) had at a temeprature 200°C higher. Their surface scales were generally smooth, dull, and very thin, as shown in Figure 108. The Coors material, however, exhibited some crystalline portions of the scale, as shown in Figure 109.

The most interesting aspect of Si/SiC materials is their retained strength after exposure. Table 41 detailed and Table 43 summarized the residual strength data. It is observed that a range from a 30% strength increase to a 50% strength loss was observed for these materials. This large difference and discrepancy when compared to the retained strength of the other silicon carbide materials must be related to the silicon phase in these materials. Strength data are difficult to analyze in Si/SiC materials, since fracture source identification is usually inconclusive due to the heterogeneous nature of the two-phase material.

The Coors 1979 SC-1 Si/SiC appears to exhibit anomalous oxidation behavior when compared to the Refel and Norton NC-430 materials. The weight gain for Coors material was higher, and it was the only material to exhibit a strength loss after exposure. This behavior is probably somehow related to the specific nature of the silicon phase in this material. A few large crystalline features on the surface of exposed samples were associated with what appeared to be silicon metal that had exuded from within the structure. Figure 110 illustrates a macroscopic crack that spontaneously developed in the Norton NC-430 material during exposure. The damaged area appears to be associated with a band of silicon-rich material that has exuded from within the composite

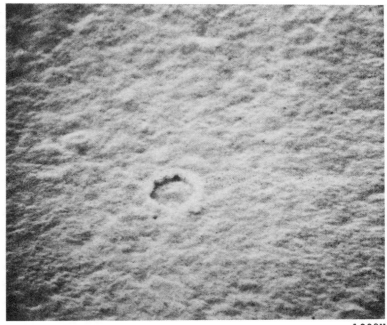

1000X

Figure 108. SEM view of surface of Refel Si/SiC after 1000 hr/1200°C static air exposure.

Materials Evaluation 247

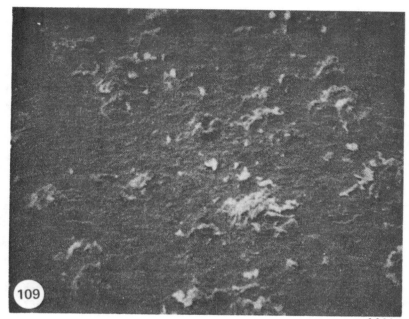

1000X

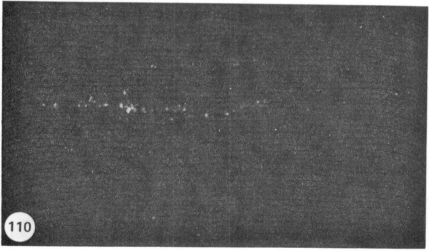

12X

Figure 109. SEM view of surface of Coors 1979 SC-1 Si/SiC after 1000 hr/1200°C static air exposure.

Figure 110. Surface of Norton NC-430 Si/SiC Sample after 1000 hr/1200°C static air exposure.

body. Behavior similar to this could certainly explain the large strength loss for the exposed Coors material.

11.3 SUMMARY OF OXIDATION IN SILICON CERAMICS

Silicon carbide materials are, in general, more stable in long term high-temperature applications than are Si_3N_4 materials. The strength distribution in SiC appears to be altered less by environmental effects such as oxidation. Weight gain by oxidation is lower in SiC, and the oxide scale thickness is less when compared to Si_3N_4. This is the case because, in general, SiC materials are of higher purity and higher density than Si_3N_4 materials. Oxidation in HP-Si_3N_4 is mainly affected by the alkali impurities that segregate in grain boundaries and then migrate to and modify the oxide scale causing increased rates of oxidation. Reaction-sintered Si_3N_4 is admittedly very pure, but a trade-off is made; the porosity is high (10-20%) in RS-Si_3N_4, and thus surface area and total oxidation are large. Sintered SiC is both pure and dense, and thus has excellent oxidation resistance. Recent forms of HP-Si_3N_4, however, show much promise for good oxidation resistance. A particular example is a Westinghouse HP-Si_3N_4 that contains nominally 4% Y_2O_3, and an undetermined amount of SiO_2.

12. Zirconia Ceramics for Diesel Engines

Zirconia ceramics have considerable promise for use in diesel engines. The following sections outline their promise, the concept of transformation-toughening, and properties of various commercial varieties. It is important to recognize that transformation-toughened zirconia is essentially a low-temperature material. It will be demonstrated below that extended high-temperature exposure can lead to overaging with resultant loss of properties.

12.1 THE ADVANTAGES OF ZIRCONIA IN DIESEL ENGINES

Zirconia is currently the ceramic material that has been the most successful in the U.S. Army TACOM Adiabatic Diesel Engine program at Cummins Engine Company.[92-95] The concept of that program is to insulate the high-temperature components of the engine such as the piston (cap), cylinder head, valve hardware, cylinder liner, and exhaust ports. Additional power and improved efficiency result from this concept since thermal energy normally lost to the cooling water and exhaust gas (almost two-thirds of the energy input in a conventional diesel engine) is converted to useful power through the use of turbocompounding. By reducing the lost energy and eliminating the need for a conventional water-cooling system, the adiabatic diesel has been demonstrated to improve fuel economy and increase power output.

The performance advances that have been achieved by the Adiabatic Diesel Engine have largely been realized through the use of one generic class of ceramic material, transformation-toughened partially stabilized zirconia.[96,97] This particular ceramic has been successful in this application for three reasons: (a) it is highly refractory; (b) it possesses a very low thermal conductivity; and (c) significant increase in fracture toughness has been attained in this ceramic material by a phenomenon known as phase transformation-toughening. Additionally, zirconia has a high thermal expansion (for ceramics), almost as high as cast iron. This means that the shrink-fit of a zirconia liner in a cast iron block will be maintained at elevated temperatures.

Much work is currently being done within the ceramics industry on transformation-toughened zirconia (TTZ). Various manufacturers are beginning to supply newly developed materials to various engine demonstration and component development programs. It is now an appropriate time to develop a data base for such zirconia ceramics.

12.2 POLYMORPHISM AND PHASE STABILIZATION IN ZIRCONIA

Pure zirconia (i.e., nonstabilized) exhibits three crystal structures between room-temperature and its melting point. Zirconia is normally monoclinic up to ~1100°C, where it transforms to tetragonal symmetry, followed by a final transformation to the cubic structure at 2370°C. The problem with pure ZrO_2 arises upon cooling through the martensitic tetragonal-to-monoclinic transformation at ~1100°C. There, approximately a 3% volume expansion occurs leading to extensive macrocracking in pure zirconia bodies.

This problem can be avoided by doping the zirconia with additions of CaO, MgO, or Y_2O_3. Thus, the material is stabilized to its high-temperature form, the cubic crystal structure. This material is termed fully stabilized zirconia (FSZ). FSZ can be cycled from room temperature to its melting point without any destructive phase transformations. Yttria has been found to be the best stabilizer, since it is less volatile than calcium and magnesium compounds at elevated temperature. Thus, destabilization of the FSZ body is less likely to occur. However, fully stabilized zirconia has disadvantages in structural applications. It is generally a coarse-grained structure, of fairly low strength, and not very tough. Its biggest drawback, perhaps, is its high thermal expansion coefficient and resulting poor thermal shock resistance. However, it is noted that the low thermal conductivity of zirconia, which is an advantage for cylinder liner application, is also a major contribution to poor thermal shock resistance in zirconia.

If less stabilizer (CaO, MgO, Y_2O_3) is incorporated into the structure than is required to fully retain the ZrO_2 cubic structure, the material is termed partially stabilized. It has long been known that this has distinct advantages, since the thermal expansion of PSZ is lower than that of FSZ, and thus PSZ has better thermal shock resistance. However, thermal shock resistance is still the performance-limiting factor of PSZ in many engineering applications.

Partially stabilized zirconia typically consists of cubic zirconia grains with second phase tetragonal and/or monoclinic precipitates. Recall that the stable structure at room temperature is monoclinic, and that a large shape and volume change is associated with the martensitic tetragonal-to-monoclinic transformation at ~1100°C. Through appropriate processing, it has recently been demonstrated how the tetragonal precipitates in PSZ can be held (in a metastable condition) within the cubic matrix at room temperature. It was found that they would revert to their stable monoclinic structure with the application of a certain stress field.

Thus originated the partially stabilized zirconia that is transformation-toughened (TTZ). Such TTZ bodies show much improved strength and toughness (about a factor of three), and are responsible for the birth (or rebirth) of zirconia as a structural engineering material, toughness being a central issue in the utilization of ceramics in structural applications.

12.3 CONCEPT OF TRANSFORMATION-TOUGHENING

The most mature form of transformation-toughened zirconia (TTZ) consists of precipitates of metastable tetragonal zirconia in a magnesia-stabilized cubic zirconia matrix.[98-113] The tetragonal precipitates are dispersed within much larger cubic grains. The tetragonal precipitates that are within a certain size range are constrained from transforming to the monoclinic structure by the surrounding matrix. However, under an applied stress the transformation occurs. The stress-induced tetragonal-to-monoclinic phase transformation is accompanied by a volume expansion. This results in the generation of stress fields around the transformed region and microcrack generation. Such two-phase transformation-toughened materials exhibit dramatic increases in strength, fracture energy, and fracture toughness. This effect has been observed in ZrO_2, Al_2O_3, and Si_3N_4 matrices, all with a dispersed ZrO_2 (tetragonal) phase.

Current mechanistic interpretations of the improved fracture properties of transformation-toughened zirconia include: (1) deflection of the advancing crack front by interation with the stress fields around the transformed areas; and/or (c) microcrack generation leading to crack branching and an increase in the energy necessary to continue crack propagation. A third mechanism is sometimes discussed as a minor contributor, i.e., energy absorption by the phase transformation process itself. Whatever the mechanism, it is known that the stress-induced phase transformation in front of an advancing crack tip does absorb energy, arrest crack propagation, and thereby increase strength and toughness.

It is important to understand that the microstructure and properties of TTZ are strongly dependent on the particle size of the initial powder, the concentration of stabilizing oxide, the time-temperature processing schedule, post-fabrication heat treatment, and surface preparation. It is possible to fabricate dense ZrO_2 bodies with various phase assemblages (in particular, varying amounts of tetragonal phase), with various amounts of microcracking, and with a varying martensitic phase transformation temperature. Generally, TTZ is considered a low temperature material. High-temperature isothermal aging is known to cause coarsening of the tetragonal phase and loss of toughness.[98] The specific time-temperature conditions that result in coarsening depend on the amount and type of rare-earth oxide used, the initial particle size, and the nature of the overall phase assemblage, etc.

12.4 EVALUATION OF VARIOUS COMMERCIAL ZIRCONIA MATERIALS

Various foreign and domestic zirconia ceramics were characterized. They were supplied to this program by R. Kamo of Cummins Engine Company. Investigated were crystal phases, stabilization, microstructure, thermal expansion, strength, elastic modulus, and fracture mode. Interpretation is made with respect to transformation-toughening, phase stability, and overaging phenomena. Five different foreign and domestic transformation-toughened zirconia materials were investigated. We have used the identifying notations A, G, J, and U to designate material from Australian, German, Japanese, and domestic sources, respectively.

12.4.1 Density, Phase Assemblage

The bulk density of each material is shown in Table 44. All are in the range 5.7-5.8 g cm^{-3}. Table 45 presents results for spectrographic cation analysis. Materials A-1, A-2, G-1, and U-4 are magnesia stabilized, whereas the Japanese material is stabilized with Y_2O_3. Note that materials A-1 and U-4 apparently contain only half the magnesia stabilizer compared to materials A-2 and G-1. Note also that the Y_2O_3-stabilized material contains a substantial amount of Al and Si. Presumably, there is a fair amount of SiO_2 at the grain boundaries. The detrimental result of this in high temperature deformation will be discussed later.

Table 46 presents the results of X-ray diffraction analysis. Note that the low MgO content in material A-1 has resulted in very little retention of the cubic phase. All other materials exhibited substantial cubic content. Additionally, A-2, J-1, and U-4 exhibited major tetragonal peaks. Recall that it is the metastable tetragonal phase that results in increased toughness in TTZ. Note also in Table 46 that the two Australian materials each contain a major monoclinic phase. Monoclinic ZrO_2 was only detected as a minor phase in the other materials.

TABLE 44. ZIRCONIA MATERIALS EVALUATED

Material Supplied[a]	Bulk Density, g cm^{-3}	Measured Open Porosity, %
A-1	5.719	0.02
A-2	5.757	0.02
G-1	5.776	0.01
J-1	5.791	0.00
U-4	5.739	0.01

[a]The notations A, G, J, and U refer to Australian, German, Japanese, and domestic sources, respectively.

TABLE 45. SPECTROGRAPHIC ANALYSIS[a] OF MAJOR IMPURITIES IN ZIRCONIA CERAMICS

Material[b]	Analysis, approximate weight percent[c]
A-1	1.1 Mg, ~1.0 Hf, 0.2 Fe, 0.1 Si
A-2	3.0 Mg, 0.1 Hf, 0.07 Si
G-1	3.0 Mg, 0.1 Hf, 0.05 Cr
J-1	(P)Y, 1.0 Al, 1.0 Si, 0.1 Hf, 0.05 Mg, 0.05 Ti
U-4	0.9-1.5 Mg

[a]Performed at AFWAL.

[b]The A, G, J, and U notations refer to Australian, German, Japanese, and domestic sources, respectively.

[c]The P notation refers to a primary constituent (e.g., Zr); it is used here to indicate two materials that are yttria-stabilized (i.e., Y-PSZ).

TABLE 46. X-RAY DIFFRACTION ANALYSIS RESULTS[a]

Material[b]	Phases Present[c]	
	Major	Minor
A-1	tet., mono.	
A-2	cubic, tet., mono.	
G-1	cubic	tet., mono.
J-1	cubic, tet.	mono. (trace)
U-4	tet., cubic	mono.

[a] Performed at AFWAL/ML.

[b] A. G. J. and U notations refer to Australian, German, Japanese, and domestic sources, respectively.

[c] The notations tet and mono. refer to the tetragonal and monoclinic crystal structures, respectively.

12.4.2 Microstructure

The microstructure of these materials was studied by reflected light microscopy. Materials A-1, A-2, G-1, and U-4 look similar, as shown in Figures 111-114. These are all magnesia-stabilized, and consist of large ~50 μm grains. From the XRD results, it would be inferred that these large grains were typically cubic. However, note that cubic ZrO_2 was not detected, even as a minor constituent, for material A-1. This is not understood at present, since both A-1 and A-2 exhibited total extinction when viewed in the petrographic microscope with crossed polars. This would be expected for the isotropic cubic crystal, but not for anisotropic tetragonal or monoclinic crystals. Figure 115 illustrates that in contrast to the large grain magnesia-stabilized materials, the Y_2O_3-stabilized Japanese material (J-1) is very fine grained (~1-5 μm). Some submicron, presumably tetragonal, particles are also observed in the SEM micrograph of the fracture surface of J-1 shown in Figure 115.

12.4.3 Strength, Deformation, and Fracture

The strength of these five transformation-toughened materials is shown in Figure 116. Tests were generally conducted in four point flexure at 25° and 1000°C. Note that the fine grain J-1 material has the highest strength, ~105 ksi at 25°C. This shows the promise of Y_2O_3-stabilized material. However, it is demonstrated that this fine grained Y_2O_3-stabilized material deforms more at elevated temperature than the MgO-stabilized materials. This appears to be a long-term creep effect rather than a short-time effect. It will be shown below that the J-1 material deformed badly during the thermal expansion test. This may be attributed to enhanced grain boundary creep due to the fine grain size of this material. Also, Table 45 illustrated that the J-1 material had substantial Si impurity, which is usually associated with SiO_2 in intergranular regions, resulting in large high temperature deformation. However, at short times this is not as obvious. Figure 117 illustrates that the J-1 material exhibited linear stress strain behavior at 1000°C. This would

Figure 111. Microstructure of zirconia A-1 polished and etched 50 sec.

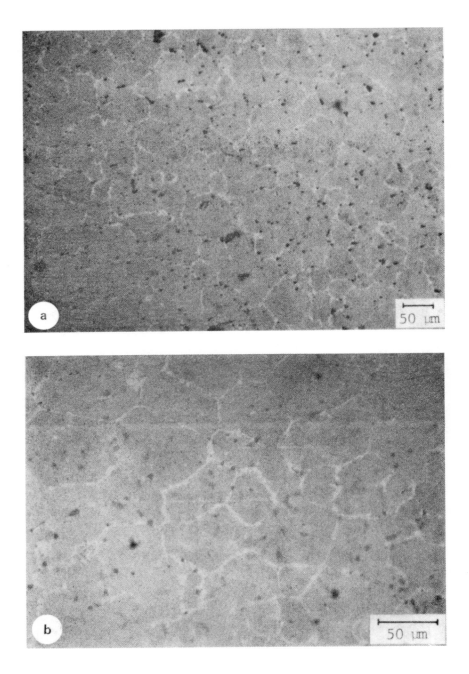

Figure 112. Microstructure of zirconia A-2 polished and etched 30 sec.

Materials Evaluation 259

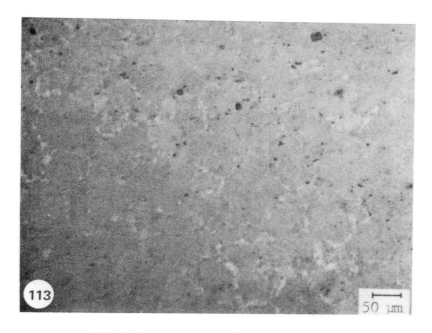

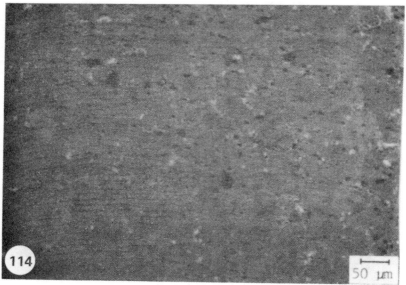

Figure 113. Microstructure of zirconia G-1 polished and etched 50 sec.

Figure 114. Microstructure of zirconia U-4 polished and etched 55 sec.

260 Ceramic Materials for Advanced Heat Engines

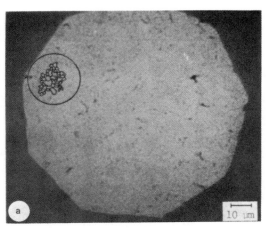

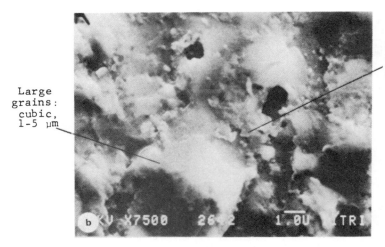

Figure 115. Microstructure of zirconia J-1.
(a) Polished and etched 35 sec.
(b) SEM micrograph of the fracture surface showing grain morphology.

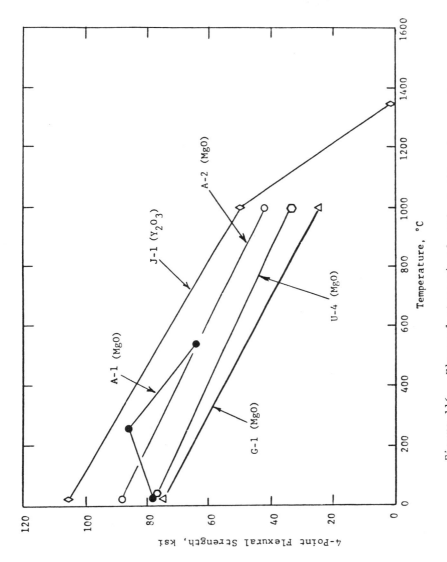

Figure 116. Flexural strength of various zirconia ceramics.

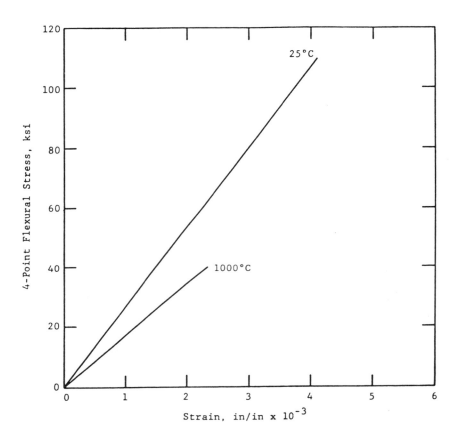

Figure 117. Representative flexural stress-strain behavior of zirconia J-1.

tend to rule out deformable grain boundary phases. However, Table 47 illustrates that strain-to-failure of the Y_2O_3-doped J-1 material is higher than that of the MgO-stabilized materials. Part of this effect, however, is due to the fact that the J-1 material was slightly stronger than the others (Figure 116).

TABLE 47. DEFORMATION OF TTZ AT 1000°C

Material	Failure Strain, 10^{-3}	Elastic Modulus, 10^6 psi
A-2	1.65	24.3
G-1	1.54	23.7
U-4	1.48	22.2
J-1	1.9, 3.0	17.1, 25.2

The two Australian materials are nominally 85 ksi materials. They are judged to be the most mature TTZ today. Much of the technology originated at CSIRO.

There are several domestic organizations developing TTZ, and it is encouraging to see a domestic ceramic supplier producing a high quality technical grade tough zirconia material. The strength of the U-4 material was 75-80 ksi.

The fracture mode appears mainly transgranular in these transformation-toughened ZrO_2 materials, as shown in Figures 118-125. A large inclusion or particle with larger thermal expansion than surrounding material is shown in Figure 118. Porosity is observed on the fracture surface of A-2 in Figure 119.

The fracture surface of G-1 shown in Figure 120 is especially interesting. Note the large grains, and intra- as well as intergranular porosity. The fracture mode in this material is entirely transgranular. This fracture surface is very similar to the TTZ fracture surface shown by Rice et al.,[99] where fracture origins were found to be predominantly single grain-boundary facets (i.e., the portion of a boundary between two grains, as opposed to grain junction triple points). Such intergranular

264 *Ceramic Materials for Advanced Heat Engines*

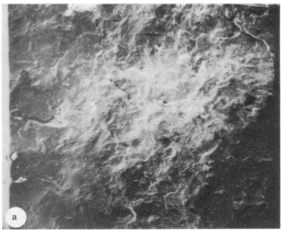

ZT2F1

ZT2F3

Figure 118. SEM fracture surfaces of zirconia A-2.

(a) Far from fracture origin on sample broken at 25°C (90.6 ksi)

(b) Fracture origin on sample broken at 1000°C (tensile surfaces together, 38.5 ksi)

Materials Evaluation 265

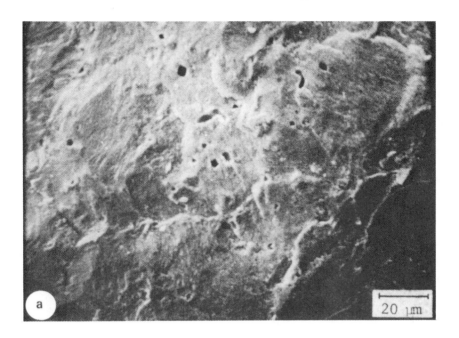

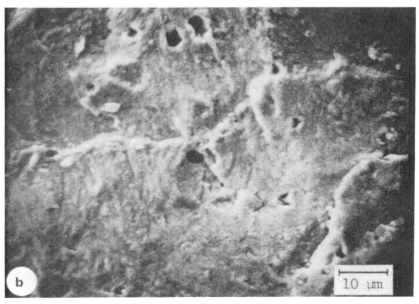

Figure 119. SEM micrographs of zirconia A-2 broken at 25°C (far from the tensile surface).

ZT3F1

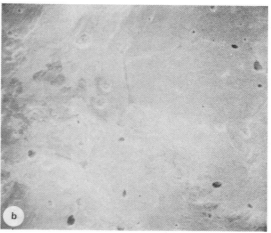

Zt3F4

Figure 120. SEM fracture surfaces of zirconia G-1.

(a) Far from origin on sample broken at 25°C (tensile surface is at bottom), 75.2 ksi

(b) Far from origin on sample broken at 1000°C (note large grains). 37.6 ksi

fracture origins completely surrounded by transgranular failure also seem to be the case for the G-1 material. This fracture mode for the G-1 material is consistent with Lange's observations[113] that a cubic material produces a smooth, transgranular fracture (note in Table 46 that G-1 is predominantly cubic).

The fracture mode of the domestic U-4 material also appears transgranular, as shown in Figures 121-123. Intergranular porosity and large grain fracture origins are illustrated.

The J-1 material has a much finer grain size, and its fracture surfaces are shown in Figures 124 and 125. Note the high magnification of these photographs. The fracture mode appears to be irregular transgranular, which would correlate with Lange's observations[113] of the fracture mode in TTZ when substantial tetragonal phase was present. Note in Figure 125 a region of poor bonding and high local porosity in the J-1 material (subsurface, at the center of the specimen width).

12.4.4 Thermal Expansion, Long-Term Stability

Thermal expansion was evaluated using an automatic recording pushrod dilatometer. Measurements were made to 1500°C at a very slow rate, 1°C min^{-1}. It is recognized that these materials were not necessarily designed to be used at temperatures as high as 1500°C; however, testing to that temperature provides information that leads to a better understanding of microstructure, stability, role of impurities, etc., in PSZ.

The thermal expansion results are shown in Figures 126-129 and summarized in Table 48. It is observed that the A-2, G-1, and U-4 materials all exhibited stable behavior upon heating, but that a distinct indication of a phase transformation is apparent upon cooling. This phenomenon is similar to that described by Hannink et al.[98] when investigating isothermal aging of TTZ. If the tetragonal precipitate phase is within a certain small size range, then the matrix constrains the particles, the tetragonal phase is retained, and the transformation-toughening mechanism is

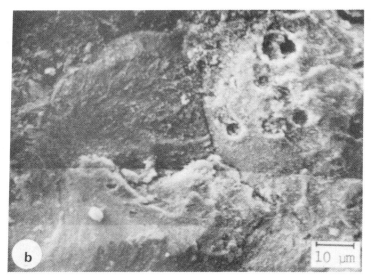

Figure 121. Fracture surfaces of zirconia U-4.
(a) SEM micrograph of 25°C fracture surface
(b) SEM micrograph of sample broken at 25°C (far from the tensile surface).

Figure 122. SEM fracture surface of zirconia U-4 at 1000°C, illustrating a large grain as the fracture origin.

Figure 123. Fracture surfaces of zirconia U-4.

(a) SEM micrograph of fracture surfaces of zirconia U-4 broken at 1000°C in flexure (tensile surfaces mating).

(b) Higher magnification of fracture surface illustrated in (a) above.

Materials Evaluation 271

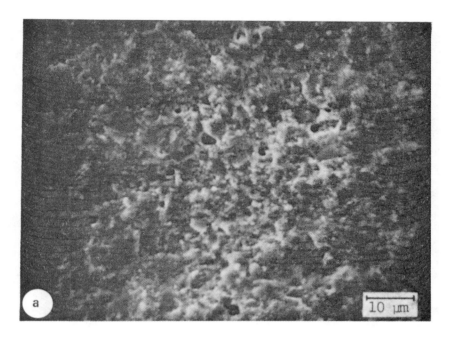

Figure 124. SEM micrographs of zirconia J-1 broken at 25°C (far from the tensile surface.

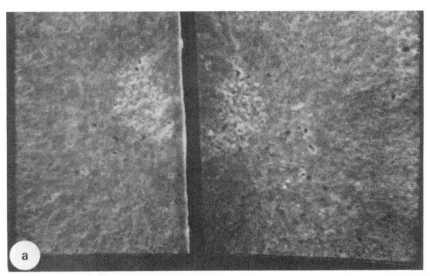

475X

950X

Figure 125. SEM fracture surfaces of zirconia J-1
(a) Sample broken at 1000°C (tensile faces together) illustrating a region of poor bonding and high local porosity (51.6 ksi)
(b) Higher magnification of (a) above.

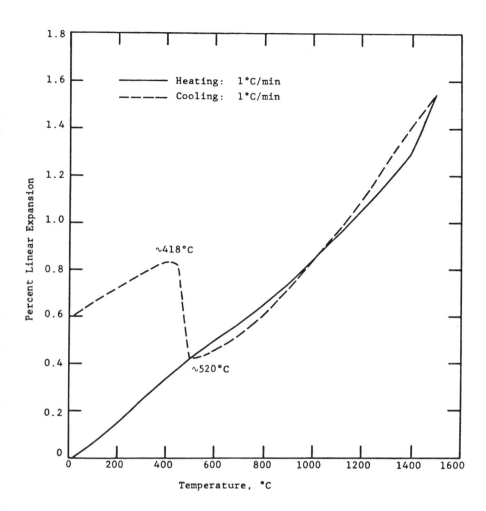

Figure 126. Thermal expansion of zirconia A-2.

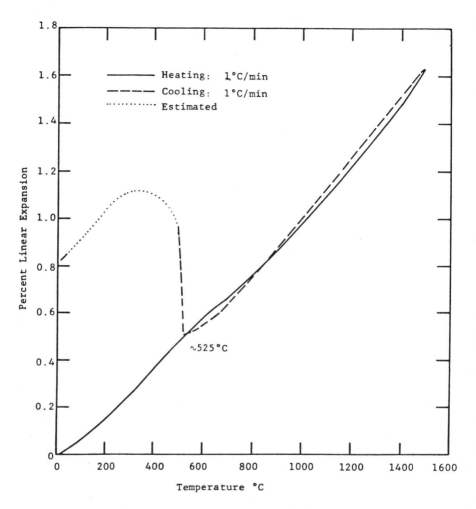

Figure 127. Thermal expansion of zirconia G-1.

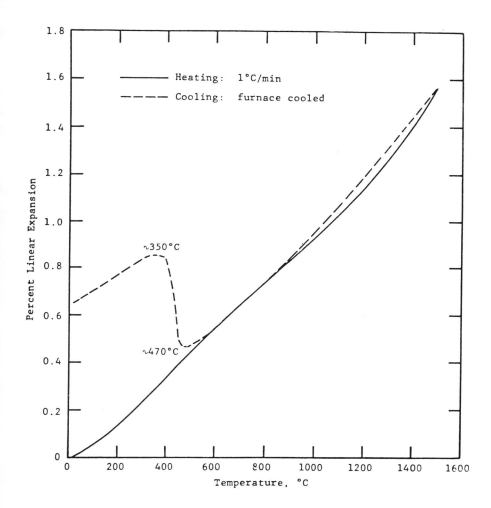

Figure 128. Thermal expansion of zirconia U-4.

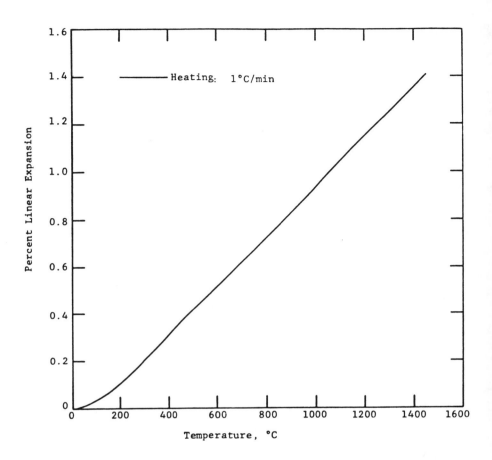

Figure 129. Thermal expansion of zirconia J-1 (sample began to deform plasticially after 1200°C).

TABLE 48. THERMAL EXPANSION OF ZIRCONIA

Material	Percent Linear Expansion at 1000°C	
	Heating	Cooling
A-2	0.84	0.85
G-1	0.98	1.0
J-1	0.94	--
U-4	0.93	0.96

operative. Magnesia- and calcia-stabilized structures promote the coarsening of tetragonal precipitates during prolonged isothermal aging. Such large tetragonal particles will transform to the monoclinic symmetry upon subsequent cooling to room-temperature. X-ray analysis of these samples after thermal exposure confirms this. Table 49 illlustrates that after the one cycle expansion test to 1500°C, the exposed A-2, G-1, and U-4 magnesia-stabilized materials contained much transformed monoclinic phase. The metastable tetragonal phase was not retained. Note that the microcracking associated with this overaging phenomenon resulted in a greater than 60% strength decrease for these materials.

The fine grain Y_2O_3-stabilized J-1 material is particularly interesting. Figure 129 illustrates stable heating cycle thermal expansion behavior. However, this sample deformed badly (plastically) at $T > 1200°C$, and the cooling cycle expansion record could not be completed. However, Table 49 illustrates that the tetragonal phase was retained after the thermal exposure. Monoclinic zirconia was not detected in the exposed sample. Therefore, this demonstrates that the overaging phenomenon is much less pronounced in the Y_2O_3-stabilized material. Consistent with this observation, as shown in Table 49, is that the Y_2O_3-stabilized J-1 material exhibited only a 28% strength reduction after exposure. This is compared to the 60-70% strength loss for the MgO-stabilized materials.

TABLE 49. RESIDUAL STRENGTH AND PHASE ASSEMBLAGE OF ZIRCONIA MATERIALS AFTER THERMAL CYCLING[a]

Material	Bulk Density, g cm⁻³	Room Temperature 4-Point Flexure Strength, ksi		X-Ray Phase Identification[b]	
		As-Received	After One Thermal Cycle	As-Received	After One Thermal Cycle
G-1 (MgO)	5.776	75.0	29.0 (-61%)	cubic	mono., cubic (tet.?)
U-4 (MgO)	5.739	77.3	24.9 (-68%)	tet., cubic	mono., cubic (tet.?)
A-2 (MgO)	5.757	88.5	--	cubic, tet., mono.	mono., cubic
J-1 (Y₂O₃)	5.791	105.0	75.7 (-28%)	cubic, tet.	cubic, tet.

[a] Thermal cycle consisted of heating to 1500°C followed by cooling to 25°C, both at 1°C/min.

[b] Major phases only, conducted on a polished cross-section, believed to be representative of the bulk.

12.4.5 Conclusion

A domestic supplier is producing a transformation-toughened zirconia that is comparable in microstructure and properties to more mature foreign materials. Y_2O_3-stabilized materials have much finer grain size than MgO-stabilized materials. This results in superior (~30%) room temperature strength. However, creep may be a problem at elevated temperature with the Y_2O_3-stabilized materials. The fine grain Y_2O_3 systems appear to have less severe overaging problems associated with long-term high-temperature isothermal exposure when compared with the MgO-stabilized ZrO_2.

13. Concluding Remarks

13.1 SILICON-BASE CERAMICS

A comparison of the major characteristics of the dense forms of SiC and Si_3N_4 is summarized in Table 50. SiC is intermediate in strength to Si_3N_4. However, SiC experiences much less strength degradation by subcritical crack growth, and thus SiC is superior at temperatures greater than 1400°C. Similarly, SiC is more stable and oxidation-resistant at extremely high temperatures. It is, in general, a purer material and not as affected by impurities or secondary phases as Si_3N_4 is. SiC has extremely good creep strength, which coupled with its good oxidation resistance and microstructural stability, make SiC a likely candidate for very long time, high temperature applications. SiC has much higher thermal conductivity and thermal diffusivity compared to Si_3N_4. This means that SiC would be a good candidate for a combustor in a gas turbine engine, but would not be a good candidate as a cast iron diesel engine cylinder liner.

The major disadvantages of SiC when directly compared to Si_3N_4 are its lower fracture toughness and lower thermal shock resistance. The low toughness of SiC is due to its low critical stress intensity factor and low fracture surface energy. Poor thermal shock resistance is often cited as the most critical difference between silicon carbide and silicon nitride. The low thermal shock resistance of SiC is due to the combination of its higher thermal expansion and higher elastic modulus in comparison to Si_3N_4. These properties are more or less inherent, and cannot be modified to any meaningful extent by varying composition or processing method. Thus SiC and Si_3N_4 are unique as engineering materials as are the various heat engine component designs for which they are being used.

During the time frame of this program, the structural ceramic technology can be characterized as a period when researchers

TABLE 50. A COMPARISON OF MAJOR CHARACTERISTICS OF DENSE FORMS
OF SILICON CARBIDE AND SILICON NITRIDE

Silicon Carbide	Silicon Nitride
• Intermediate Strength	• High Strength
• High Thermal Expansion	• Low Thermal Expansion
• Poor Thermal Shock Resistance	• Good Thermal Shock Resistance
• Low Fracture Toughness	• Higher Fracture Toughness
• High Temperature Stability	• Strength Degradation by Slow-Crack Growth
• High Thermal Conductivity	• Low Thermal Conductivity
• Good Oxidation Resistance	• Oxidation Strong Function of Impurities
• High Creep Strength	• Low Creep Strength
• Low Influence of Impurities	• Second Phases Influence Behavior
• High Elastic Modulus	• Lower Elastic Modulus
• Transgranular Fracture	• Intergranular Fracture
• Little Subcritical Crack Growth	• Nonlinear Stress-Strain

realized the inadequacy of magnesia additives for HP-Si_3N_4, and the tendency of some Y_2O_3 compositions to form unstable phases. These problems have been solved. We have demonsterated that Y_2O_3- and ZrO_2-modified HP-Si_3N_4 materials are being produced that have excellent elevated temperature properties.

Also during the current reporting period, the technology was developed to produce sintered silicon carbide in a variety of complex component configurations. However, based on the sintered silicon carbide materials we have evaluated on this program and its predecessor, it appears that advances in SiC technology are not being made as rapidly as advances in Si_3N_4 technology. For Si_3N_4, the advances are being made by improvement of the performance-limiting aspect of that material--the intergranular phase, and its deformation at elevated temperature. For SiC the critical issues appear to be the elimination of isolated surface-connected porosity, and the inhibition of exaggerated α-phase grain growth. It may be that these performance-limiting aspects are more intrinsic for silicon carbide; perhaps SiC is nearer its inherent performance limit. The elimination of viscous intergranular regions in Si_3N_4 may be a more tractable problem.

13.2 TRANSFORMATION-TOUGHENED ZIRCONIA

Transformation-toughened zirconia is being considered for structural use in the hot sections of diesel engines. Its main attributes are low thermal conductivity and thermal expansion high enough to effectively used with cast iron engine components. It has a demonstrated high fracture toughness when compared with silicon-base ceramics. However, it must be recognized that TTZ is essentially a low-temperature material. Overaging phenomena, resulting in loss of toughness and strength, will occur during extended isothermal exposure above the martensitic transformation temperature (~900°C). Those unfamiliar with the transformation toughening mechanism should use this material with caution. Partially stabilized zirconia has long been used in high-temperature refractory applications. The same high temperature stability does not exist for transformation-toughened zirconia, however.

References

1. D. C. Larsen, "Property Screening and Evaluation of Ceramic Turbine Engine Materials," AFML-TR-79-4188, October, 1979.

2. D. C. Larsen and J. W. Adams, "Property Screening and Evaluation of Ceramic Turbine Materials," IITRI Semiannual Interim Report No. 8, AFWAL Contract F33615-79-C-5100, June 1980.

3. D. C. Larsen and J. W. Adams, "Property Screening and Evaluation of Ceramic Turbine Materials," IITRI Semiannual Interim Technical Report No. 9, AFWAL Contract F33615-79-C-5100, December 1980.

4. D. C. Larsen and J. W. Adams, "Property Screening and Evaluation of Ceramic Turbine Materials," IITRI Semiannual Interim Technical Report No. 10, AFWAL Contract No. F33615-79-C-5100, April 1981.

5. D. C. Larsen and J. W. Adams, "Property Screening and Evaluation of Ceramic Turbine Materials," IITRI Semiannual Interim Technical Report No. 11, AFWAL Contract F33615-79-C-5100, November 1981.

6. D. C. Larsen and J. W. Adams, "Property Screening and Evaluation of Ceramic Turbine Materials," IITRI Semiannual Interim Technical Report No. 12, AFWAL Contract F33615-79-C-5100, May 1982.

7. S. A. Bortz and D. C. Larsen, "Properties of Structural Ceramics," paper presented at 4th Annual Conference on Materials for Coal Conversion and Utilization, U.S. National Bureau of Standards, Gaithersburg, Maryland, October 9-11, 1979.

8. D. C. Larsen and R. Ruh, "Thermal Diffusivity of Silicon-Base Ceramics for Gas Turbine Applications," paper presented at the 16th International Thermal Conductivity Conference, IIT Research Institute, Chicago, Illinois, November 7-9, 1979.

9. D. C. Larsen, H. H. Nakamura, and R. Ruh, "Thermal Expansion of Silicon-Base Gas Turbine Ceramics," paper presented at the 7th International Thermal Expansion Symposium, IIT Research Institute, Chicago, Illinois, November 7-9, 1979.

10. M. G. Mendiratta, P. L. Land, R. Ruh, R. W. Rice, and D. C. Larsen, "Fractography of Reaction-Sintered Si_3N_4," Trans. ASME, J. Engr. Power, $\underline{102}$, pp. 244-248, 1980.

REFERENCES (cont.)

11. S. A. Bortz and D. C. Larsen, "Properties of Structural Ceramics," SAMPE Journal, January/February 1981.

12. D. C. Larsen, J. W. Adams, and R. Ruh, "Corrosion of Silicon-Ceramics and Oxide Ceramics in Coal Gas Environment," paper presented at the American Ceramic Society 83rd Annual Meeting, Washington, D.C., May 3-6, 1981.

13. D. C. Larsen, J. W. Adams, and R. Ruh, "Oxidation of Silicon Nitride and Silicon Carbide," paper presented at American Ceramic Society 83rd Annual Meeting, Washington, D.C., May 3-6, 1981.

14. D. C. Larsen, J. W. Adams, S. A. Bortz, and R. Ruh, "Evidence of Strength Degradation by Subcritical Crack Growth in Si_3N_4 and SiC," presented at the International Symposium on Fracture Mechanics of Ceramics, July 15-17, 1981, Pennsylvania State University, University Park, PA, in Fracture Mechanics of Ceramics, Vol. 5, R. C. Bradt, A. G. Evans, D. P. H. Hasselman, and F. F. Lange (eds.), pp. 571-586, Plenum Press, New York (1983).

15. D. C. Larsen, J. W. Adams, and R. Ruh, "The Nature of SiC for Use in Heat Engines Compared to Si_3N_4: An Overview of Property Differences," paper presented at NATO Advanced Study Institute on Nitrogen Ceramics, University of Sussex, Brighton, U.K., 27 July-7 August 1981, in Progress in Nitrogen Ceramics, F. L. Riley (ed.), pp. 695-710, Martinus Nijhoff Publishers, Boston, Massachusetts (1983).

16. J. W. Adams, D. C. Larsen, and R. Ruh, "Strength-Microstructure Relations for Various SiC Ceramics," paper presented at American Ceramic Society Annual Meeting, Cincinnati, OH, May 1982.

17. D. C. Larsen, J. W. Adams, and R. Ruh, "Mechanical and Thermal Characterization of Foreign and Domestic Zirconia Ceramics for Diesel Engines," presented at the 85th Annual Meeting of the American Ceramic Society, Chicago, April 24-27, 1983.

18. J. W. Adams, D. C. Larsen, and R. Ruh, "Microstructural Study of Various Silicon Nitride Ceramics," presented at the 85th Annual Meeting of the American Ceramic Society, Chicago, April 24-27, 1983.

19. C. A. Johnson and S. Prochazka, "Microstructures of Sintered SiC," in Ceramic Microstructures '76, R. M. Fulrath and J. A. Pask (eds.), Westview Press, 1977.

REFERENCES (cont.)

20. G. W. Robinson and R. E. Gardner, "Ceramographic Preparation of Silicon Carbide," J. Amer. Ceram. Soc., <u>47</u> (4), p. 201, 1964.

21. K. S. Mazdiyasni and C. M. Cooke, "Synthesis, Characterization, and Consolidation of Si_3N_4 Obtained from Ammonolysis of $SiCl_4$," J. Amer. Ceram. Soc., <u>56</u>, No. 12, p. 628, 1973.

22. G. Petzow, <u>Metallographic Etching</u>, Amer. Soc. for Metals, Metals Park, OH, p. 92.

23. K. S. Mazdiyasni and C. M. Cooke, "Consolidation, Microstructure, and Mechanical Properties of Si_3N_4-Doped with Rare-Earth Oxides," J. Amer. Ceram. Soc., <u>57</u>, No. 12, p. 537, 1974.

24. D. Cheever, Coors Porcelain Co., Private communication, 1982.

25. F. B. Seely and J. O. Smith, <u>Advanced Mechanics of Materials</u>, John Wiley and Sons, Inc., New York, 1957.

26. S. Spinner and W. E. Tefft, "A Method for Determining Mechanical Resonance Frequencies and for Calculating Elastic Moduli from These Frequencies," ASTM Proceedings, <u>61</u>, pp. 1221-1238, 1961.

27. G. Zener, <u>Elasticity and Anelasticity of Metals</u>, University of Chicago Press, 1948.

28. D. F. Moore, "Internal Friction," <u>Principles and Applications of Tribology</u>, Pergamon Press, New York, 1975.

29. N. F. Astbury and W. R. Davis, "Internal Friction in Ceramics," Trans. Brit. Ceram. Soc., <u>63</u>, pp. 1-18, 1964.

30. H. Kolsky, <u>Stress Waves in Solids</u>, Oxford-Clarendon Press, 1953.

31. W. J. Parker, et al., J. Appl. Phys., <u>32</u>, p. 1679, 1961.

32. J. A. Cape and G. W. Lehman, "Temperature and Finite Pulse Time Effects in the Flash Method for Measuring Thermal Diffusivity," J. Appl. Phys., <u>34</u>, pp. 1909-1913, 1963.

33. R. E. Taylor and J. A. Cape, "Finite Pulse-Time Effects in the Flash Diffusivity Technique," Appl. Phys. Letters, <u>5</u>, pp. 212-213, 1964.

REFERENCES (cont.)

34. R. E. Taylor, "Critical Review of Flash Method for Measuring Thermal Diffusivity," Report PRF-6764, by Thermophysical Properties Research Center, Purdue University, to National Science Foundation, October 15, 1973.

35. R. D. Cowan, "Pulse Method for Measuring Thermal Diffusivity at High Temperature," J. Appl. Phys., $\underline{34}$, pp. 926-927, 1963.

36. D. C. Larsen and J. W. Adams, "Property Screening and Evaluation of Ceramic Turbine Materials," IITRI Semiannual Interim Technical Report No. 14, AFWAL Contract F33615-79-C-5100, August, 1983.

37. A. G. Evans, "Structural Reliability--A Processing Dependent Phenomenon," J. Amer. Ceram. Soc., $\underline{65}$ (3), pp. 127-137, 1982.

38. R. W. Rice, "Microstructure Dependence of Mechanical Behavior of Ceramics," in <u>Treatise on Materials Science and Technology</u>, Vol. 11, Properties and Microstructure, R. K. MacCrone (ed.), Academic Press, 1977.

39. S. Prochazka and R. J. Charles, "Strength of Boron-Doped, Hot-Pressed Silicon Carbide," Bull. Amer. Ceram. Soc., $\underline{52}$ (12), pp. 885-891, 1973.

40. C. A. Johnson and S. Prochazka, "Microstructures of Sintered SiC" in <u>Ceramic Microstructures '76</u>, R. M. Fulrath and J. A. Pask (eds.), Westview Press, 1977.

41. P. Kennedy, J. V. Shennan, P. Braiden, J. McLaren, and R. Davidge, "An Assessment of the Performance of Refel Silicon Carbide under Conditions of Thermal Stress," Proc. Brit. Ceram. Soc. (22), pp. 67-87, June 1973.

42. D. C. Larsen and G. C. Walther, "Property Screening and Evaluation of Ceramic Turbine Engine Materials," IITRI Semiannual Interim Technical Report No. 6, AFML Contract No. F33165-75-C-5196, July 1978.

43. G. D. Quinn, "Review of Static Fatigue in Silicon Nitride and Silicon Carbide," presented at a Topical Meeting on Non-Oxide Ceramics, New England Section, American Ceramic Society, Cape Cod, MA, October 5-6, 1981.

44. D. C. Larsen and R. Ruh, "The Properties of SiC, Si_3N_4, and ZrO_2 for Engines, and the Potential of Ceramic Composites," 12th Automotive Materials Conference--Ceramics in Engines, University of Michigan, March 14-15, 1984.

REFERENCES (cont.)

45. R. R. Wills and M. C. Brockway, "Hot Isostatic Pressing of Silicon Base Ceramics, AFWL-TR-80-4193, January 1981.

46. A. G. Evans and S. M. Wiederhorn, "Crack Propagation and Failure Prediction in Silicon Nitride at Elevated Temperatures," J. Mater. Sci., $\underline{9}$ (2), pp. 270-278, 1974.

47. T. Vasilos, R. M. Cannon, Jr., and B. J. Wuensch, "Improving the Stress Rupture and Creep of Silicon Nitride," NASA-CR-159585, 1979.

48. G. G. Trantina and C. A. Johnson, "Subcritical Crack Growth in Boron-Doped SiC," J. Amer. Ceram. Soc., $\underline{58}$ (7-8), pp. 344-345, 1975.

49. A. G. Evans and S. M. Wiederhorn, "Crack Propagation and Failure Prediction in Silicon Nitride at Elevated Temperatures," J. Mater. Sci., $\underline{9}$ (2), pp. 270-278, 1974.

50. M. Srinivasan, R. H. Smoak, and J. A. Coppola, "Static Fatigue Resistance of Sintered Alpha SiC," Paper 10-C-79C presented at the American Ceramic Society 34d Annual Conference on Composites and Advanced Materials, Merritt Island, Florida, January 21-24, 1979.

51. K. D. McHenry and R. E. Tressler, "Subcritical Crack Growth in Silicon Carbide," J. Mater. Sci., $\underline{12}$, pp. 1272-1278, 1977.

52. M. S. Seltzer, "High Temperature Creep of Ceramics," AFML-TR-76-97, June 1976.

53. M. S. Seltzer, "High Temperature Creep of Silicon-Base Compounds," Bull. Amer. Ceram. Soc., $\underline{56}$ (4), pp. 418-423, 1977.

54. R. J. Charles, "High-Temperature Stress Rupture of Polycrystalline Ceramics," General Electric Report 77-CRD-036, May 1977.

55. R. W. Rice and W. J. McDonough, "Hot-Pressed Si_3N_4 with Zr-Based Additions," J. Amer. Ceram. Soc., $\underline{58}$ (5-6), p. 264, 1975.

56. W. J. McDonough, S. W. Freiman, P. F. Becher, and R. W. Rice, "Fabrication and Evaluation of Si_3N_4 Hot-Pressed with ZrO_2," ARPA/NAVSEA-Garrett/AiResearch Ceramic Gas Turbine Engine Program Review, Maine Maritime Academy, Castine, Maine, August 1-4, 1977.

REFERENCES (cont.)

57. S. W. Freiman, C. Wu, K. R. McKinney, and W. J. McDonough, "Effect of Oxidation on the Room Temperature Strength of Si_3N_4 Hot-Pressed with MgO or ZrO_2," ARPA/NAVSEA-Garrett/AiResearch Ceramic Gas Turbine Engine Program Review, Maine Maritime Academy, Castine, Maine, August 1-4, 1977.

58. D. C. Larsen and G. C. Walther, "Property Screening and Evaluation of Ceramic Turbine Engine Materials," IITRI Semiannual Interim Report No. 5, AFML Contract F33615-75-C-5196, January 1978.

59. S. Ud Din and P. S. Nicholson, "Creep Deformation of Reaction Sintered Silicon Nitrides," J. Amer. Ceram. Soc., 58 (11-12), pp. 500-502, 1975.

60. S. Grathwohl and F. Thummler, "Creep of Reaction-Bonded Silicon Nitride," J. Mater. Sci., 13, pp. 1177-1186, 1978.

61. J. M. Birch and B. Wilshire, "The Compression Creep Behavior of Silicon Nitride Ceramics," J. Mater. Sci., 13, pp. 2627-2636, 1978.

62. J. M. Birch et al., "The Influence of Stress Distribution on the Deformation and Fracture Behavior of Ceramic Materials Under Compression Creep Conditions," J. Mater. Sci., 11, pp. 1817-1825, 1976.

63. F. F. Lange, "Non-Elastic Deformation of Polycrystals with a Liquid Boundary Phase," in *Deformation of Ceramic Materials*, R. C. Bradt and R. E. Tressler (eds.), Plenum Press, New York, 1975.

64. T. L. Francis and R. L. Coble, "Creep of Polycrystalline Silicon Carbide," J. Amer. Ceram. Soc., 51 (2), pp. 115-116, 1968.

65. P. L. Farnsworth and R. L. Coble, "Deformation Behavior of Dense Polycrystalline SiC," J. Amer. Ceram. Soc., 49 (5), pp. 264-268, 1966.

66. D. P. H. Hasselman, "Thermal Stress Resistance Parameters for Brittle Refractory Ceramics: A Compendium," Bull. Amer. Ceram. Soc., 49, pp. 1033-1037, 1970.

67. D. C. Larsen and G. C. Walther, "Property Screening and Evaluation of Ceramic Vane Materials," IITRI Semiannual Interim Technical Report No. 4, AFML Contract F33615-75-C-5196, October 1977.

REFERENCES (cont.)

68. D. P. H. Hasselman, "Unified Theory of Thermal Shock Fracture Initiation and Crack Propagation in Brittle Ceramics," J. Amer. Ceram. Soc., **52**, pp. 600-604, 1969.

69. S. C. Singhal, "Oxidation of Silicon-Based Structural Ceramics," in <u>Properties of High Temperature Alloys (with Emphasis on Environmental Effects)</u>, Z. A. Foroulis and F. S. Pettit (eds.), Electrochemical Society, Inc., Princeton, NJ, pp. 697-712, 1977.

70. S. C. Singhal, "Oxidation of Silicon Nitride and Related Materials," in <u>Nitrogen Ceramics</u>, F. L. Riley (ed.), NATO Advanced Study Institute Series, Noordhoff International Publishing, pp. 607-626, 1977.

71. S. C. Singhal, "Oxidation and Corrosion-Erosion Behavior of Si_3N_4 and SiC," <u>Ceramics for High-Performance Applications</u>, J. J. Burke, A. E. Gorum, and R. N. Katz (eds.), Brook Hill Publishing Co., Chestnut Hill, MA pp. 533-548, 1974.

72. C. L. Quackenbush, J. T. Smith, J. Neil, "Oxidation in the $Si_3N_4-Y_2O_3-SiO_2$ System," Proceedings of DOE Automotive Technology Development Contractor Coordination Meeting, Dearborn, MI, November 11-14, 1980.

73. C. L. Quackenbush, "A Review of GTE Sintered Si_3N_4 Structural Ceramics," Proceedings of 5th International Symposium on Automotive Propulsion Systems, DOE CONF-800419, Vol. 1, Dearborn, MI, pp. 482-499, October 1980.

74. W. C. Tripp and H. C. Graham, "Oxidation of Si_3N_4 in the Range 1300° to 1500°C," J. Amer. Ceram. Soc., **59** (9-10), pp. 399-304, 1976.

75. S. C. Singhal, "Thermodynamics and Kinetics of Oxidation of Hot-Pressed Silicon Nitride," J. Mater. Sci., **11**, pp. 500-509, 1976.

76. S. C. Singhal and F. F. Lange, "Effect of Alumina Content on the Oxidation of Hot-Pressed Silicon Carbide," J. Amer. Ceram. Soc., **58** (9-10), 1975.

77. J. W. Hinze, W. C. Tripp, and H. C. Graham, "The High Temperature Oxidation of Hot-Pressed Silicon Carbide," in <u>Mass Transport Phenomena in Ceramics</u>, A. R. Cooper and A. H. Heuer (eds.), Plenum Press, 1975.

78. S. C. Singhal, "Oxidation Kinetics of Hot-Pressed Silicon Carbide," J. Mater. Sci., **11**, pp. 1246-1253, 1976.

REFERENCES (cont.)

79. S. C. Singhal, "Thermodynamic Analysis of the High-Temperature Stability of Silicon Nitride and Silicon Carbide," Ceramurgia Intl., $\underline{2}$ (3), pp. 123-130, 1976.

80. A. J. Kiehle et al., "Oxidation Behavior of Hot-Pressed Si_3N_4," J. Amer. Ceram. Soc., $\underline{58}$ (1-2), pp. 17-20, 1975.

81. F. F. Lange, "Si_3N_4-Ce_2O_3-SiO_2 Materials: Phase Relations and Strength," Ceram. Bull., $\underline{59}$ (2), p. 239, 1980.

82. C. L. Quackenbush and J. T. Smith, "Phase Effects of Si_3N_4 Containing Y_2O_3 or CeO_2: II, Oxidation," Ceram. Bull., $\underline{59}$ (5), p. 533, 1980.

83. J. P. Guha, P. Goursat, and M. Billy, "Hot-Pressing and Oxidation Behavior of Silicon Nitride with Ceria Additive," J. Amer. Ceram. Soc., $\underline{63}$ (1-2), p. 119, 1980.

84. F. F. Lange and B. I. Davis, "Development of Surface Stresses During the Oxidation of Several Si_3N_4/CeO_2 Materials," J. Amer. Ceram. Soc., $\underline{62}$ (11-12), p. 629, 1979.

85. H. Knoch and G. E. Gazza, "Carbon Impurity Effect on the Thermal Degradation of a Si_3N_4-Y_2O_3 Ceramic," AMMRC TR-79-27, May 1979.

86. S. Schuon, "Effect of W and WC on the Oxidation Resistance of Yttria-Doped Silicon Nitride," NASA Technical Memorandum 81528, 1980.

87. A. G. Evans and R. W. Davidge, "The Strength and Oxidation of Reaction-Sintered Silicon Nitride," J. Mater. Sci., $\underline{5}$, pp. 314-325, 1970.

88. R. W. Davidge, A. G. Evans, D. Gilling, and P. R. Wilyman, "Oxidation of Reaction-Sintered Silicon Nitride and Effects on Strength," **Special Ceramics 5**, P. Popper (ed.), British Ceramic Research Association, pp. 329-342, 1970.

89. S. C. Singhal and F. F. Lange, "Oxidation Behavior of Sialons," J. Amer. Ceram. Soc., $\underline{60}$ (3-4), p. 190, 1977.

90. M. H. Lewis and P. Barnard, "Oxidation Mechanisms in Si-Al-O-N Ceramics," J. Mater. Sci., $\underline{15}$, pp. 443-448, 1980.

91. N. J. Tighe, "Microstructural Aspects of Deformation and Oxidation of Magnesia-Doped Silicon Nitride," U.S. National Bureau of Standards Report NSBIR 76-1153, September 1976.

REFERENCES (cont.)

92. W. Bryzik, "Adiabatic Diesel Engine," Research Development, January 1978.

93. R. Kamo and W. Bryzik, "Adiabatic Turbocompound Engine Performance Prediction," SAE Technical Paper 780068, presented at SAE Congress and Exposition, February 1978.

94. R. Kamo, "Cycles and Performance Studies for Advanced Diesel Engines," in Ceramics for High Performance Applications - II, pp. 907-922, J. J. Burke, E. N. Lenoe, and R. N. Katz (eds.), Brook Hill Publishing Company, Chestnut Hill, MA, 1978.

95. R. Kamo and W. Bryzik, "Cummins-TARADCOM Adiabatic Turbocompound Engine Program," SAE Paper 810070, SAE International Congress and Exposition, February 1981.

96. R. N. Katz and E. M. Lenoe, "Ceramics for Diesel Engines: Preliminary Results of a Technology Assessment," AMMRC SP 81-1, October 1981.

97. M. E. Woods and I. Oda, "PSZ Ceramics for Adiabatic Engine Components," SAE Technical Paper 820429, presented at SAE International Congress and Exposition, Detroit, MI, February 1982.

98. R. H. J. Hannink, K. A. Johnston, R. T. Pascoe, and R. C. Garvie, "Microstructural Changes during Isothermal Aging of Calcia Partially Stabilized Zirconia Alloy," in Advances in Ceramics, Volume 3, Science and Technology of Zirconia, pp. 116-136, A. H. Heuer and L. W. Hobbs (eds.), Amer. Ceram. Soc., Columbus, OH, 1981.

99. R. W. Rice et al., "Grain Boundaries, Fracture, and Heat Treatment of Commercial Partially Stabilized Zirconia," Com. Amer. Ceram. Soc., C-175, December 1981.

100. P. G. Valentine et al., "Microstructure and Mechanical Properties of Bulk Yttria-Partially-Stabilized Zirconia," NASA Contractor Report CR-165402, August 1981.

101. A. G. Evans and A. H. Heuer, "Review: Transformation-Toughening in Ceramics: Martensitic Transformation in Crack-Tip Stress Fields," J. Amer. Ceram. Soc., $\underline{63}$ (5-6), pp. 241-248, 1980.

102. R. T. Pascoe and R. C. Garvie, "Surface Strengthing of Transformation-Toughened Zirconia," in Ceramic Microstructure '76, R. M. Fulrath and J. A. Pask (eds.), Westview Press, Boulder, CO, 1977.

REFERENCES (cont.)

103. T. K. Gupta, F. F. Lange, and J. H. Bechtold, "Effect of Stress-Induced Phase Transformation on the Properties of Polycrystalline Zirconia Containing Metastable Tetragonal Phase," J. Mater. Sci., 13, pp. 1464-1470, 1978.

104. R. C. Garvie, R. H. Hannink, and R. T. Pascoe, "Ceramic Steel?" Nature, 258, pp. 703-704, 1975.

105. R. C. Garvie, "Stabilization of the Tetragonal Structure in Zirconia Microcrystals," J. Phys. Chem., 82 (2), pp. 218-224, 1978.

106. A. H. Heuer, "Alloy Design in Partially Stabilized Zirconia," in Advances in Ceramics, Volume 3, Science and Technology of Zirconia, A. H. Heuer and L. W. Hobbs (eds.), pp. 98-115, Amer. Ceram. Soc., Columbus, OH, 1981.

107. N. Claussen and M. Ruhle, "Design of Transformation-Toughened Ceramics," in Advances in Ceramics, Volume 3, Science and Technology of Zirconia, A. H. Heuer and L. W. Hobbs (eds.), pp. 137-163, Amer. Ceram. Soc., Columbus, OH, 1981.

108. H. P. Kirchner et al., "Crack Branching in Transformation-Toughened Zirconia," J. Amer. Ceram. Soc., 64 (9), pp. 529-533, 1981.

109. T. K. Gupta, "Role of Stress-Induced Phase Transformation in Enhancing Strength and Toughness of Zirconia Ceramics," in Fracture Mechanics of Ceramics, Volume 4, Crack Growth and Microstructure, R. C. Bradt, D. P. H. Hasselman, and F. F. Lange (eds.), pp. 877-889, Plenum Press, NY, 1978.

110. N. Claussen and D. P. H. Hasselman, "Improvement of Thermal Shock Resistance of Brittle Structural Ceramics by a Dispersed Phase of Zirconia," in Thermal Stresses in Severe Environments, D. P. H. Hasselman and R. A. Heller (eds.), pp. 381-395, 1980.

111. R. W. Rice, "Further Discussion of 'Precipitation in Partially Stabilized Zirconia,'", J. Amer. Ceram. Soc., 60 (5-6), p. 280, 1977.

112. D. L. Porter and A. H. Heuer, Reply to "Further Discussion of 'Precipitation in Partially Stabilized Zirconia,'" J. Amer. Ceram. Soc., 60 (5-6), p. 280, 1977.

113. F. F. Lange, "Transformation Toughening. Part 3: Experimental Observations in the $ZrO_2-Y_2O_3$ System," J. Mater. Sci., 17, pp. 240-246, 1982.

Appendix

Reflected Light and SEM Micrographs of Polished Reaction-Sintered Silicon Nitride Materials

294 Ceramic Materials for Advanced Heat Engines

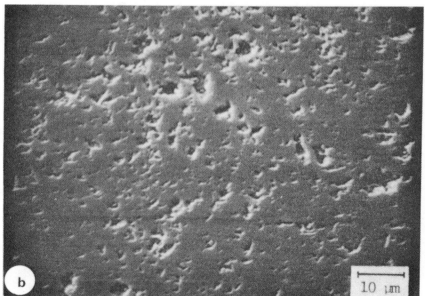

Figure A1. Micrographs of 1976 Norton NC-350 RS-Si_3N_4.
(a) Optical micrograph of polished section
(b) SEM view of polished section

Materials Evaluation 295

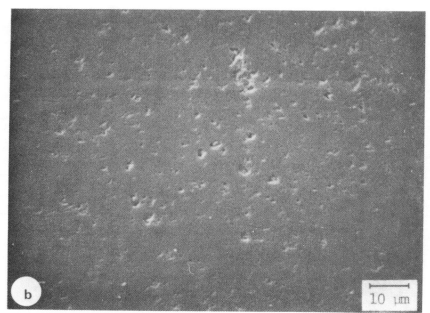

Figure A2. Micrographs of 1977 Norton NC-350 RS-Si_3N_4.
(a) Optical micrograph of polished section
(b) SEM view of polished section

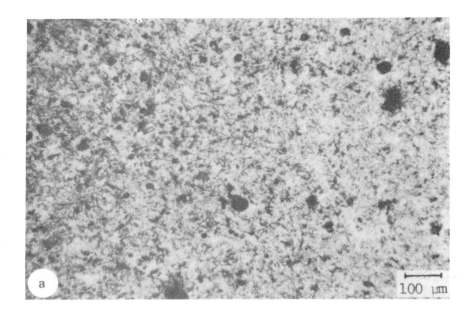

Figure A3. Micrographs of Kawecki-Berylco (batch 2) $RS-Si_3N_4$.

(a) Optical micrograph of polished section
(b) SEM view of polished section

Materials Evaluation 297

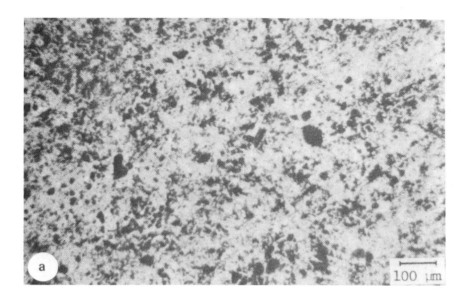

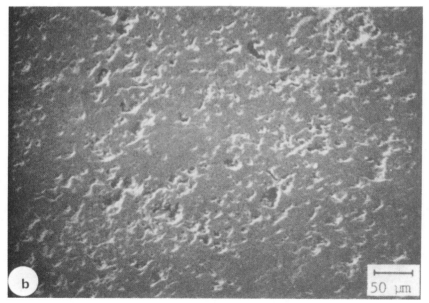

Figure A4. Micrographs of Kawecki-Berylco
(batch 3) RS-Si$_3$N$_4$.
(a) Optical micrograph of polished section
(b) SEM view of polished section

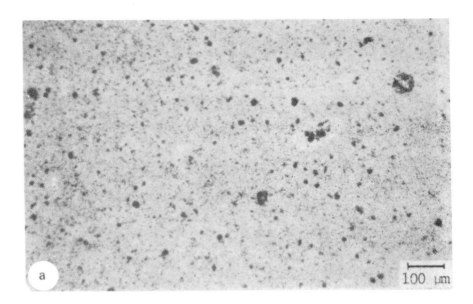

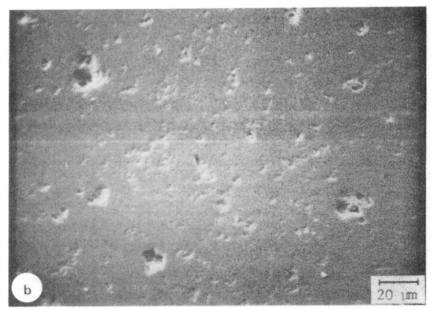

Figure A5. Micrographs of Ford injection molded RS-Si_3N_4.
(a) Optical micrograph of polished section
(b) SEM view of polished section

Materials Evaluation 299

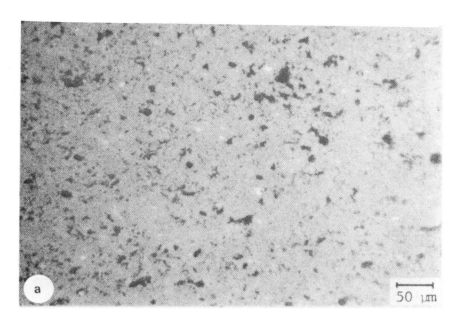

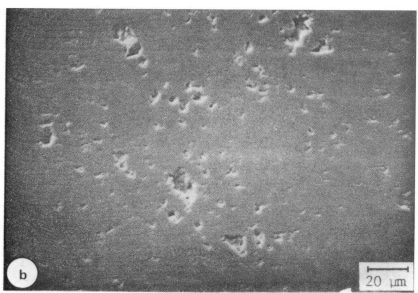

Figure A6. Micrographs of AiResearch slip cast
(Airceram RBN-101) RS-Si_3N_4.
(a) Optical micrograph of polished section
(b) SEM view of polished section

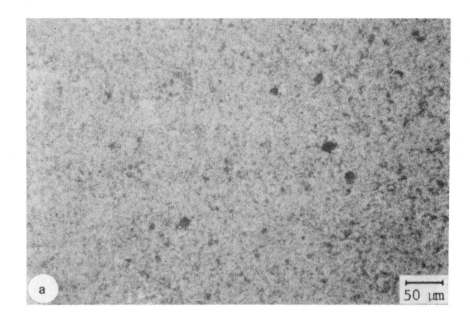

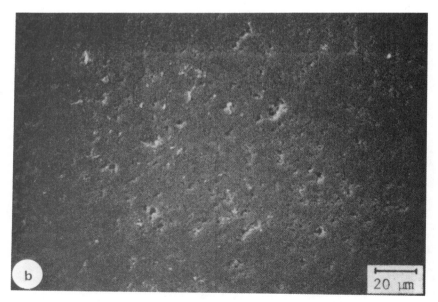

Figure A7. Micrographs of Raytheon isopressed RS-Si_3N_4.
(a) Optical micrograph of polished section
(b) SEM view of polished section

Materials Evaluation   301

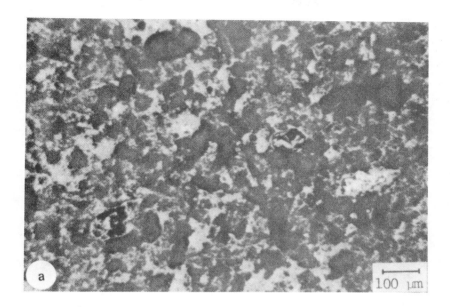

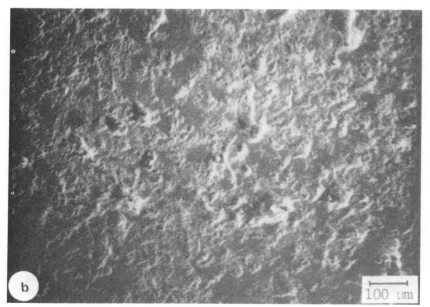

Figure A8. Micrographs of Indussa/Nippon
           Denko RS-Si_3N_4.

(a) Optical micrograph of polished section
(b) SEM view of polished section

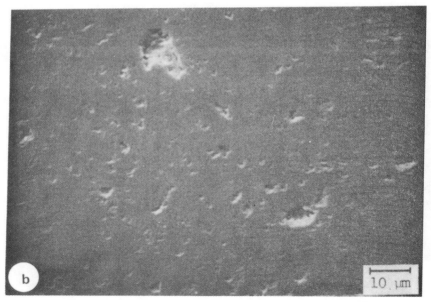

Figure A9. Micrographs of AiResearch injection molded (Airceram RBN-122) RS-Si_3N_4.

(a) Optical micrograph of polished section
(b) SEM view of polished section

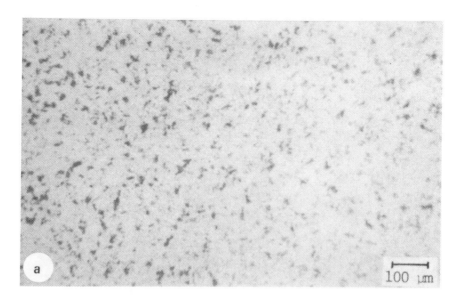

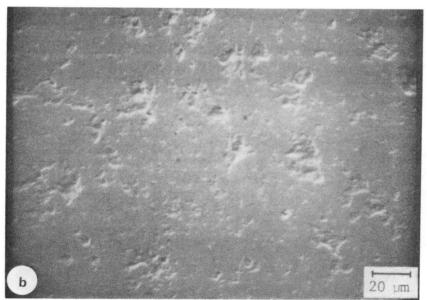

Figure A10. Micrographs of 1979 Norton NC-350 RS-Si$_3$N$_4$.

(a) Optical micrograph of polished section
(b) SEM view of polished section

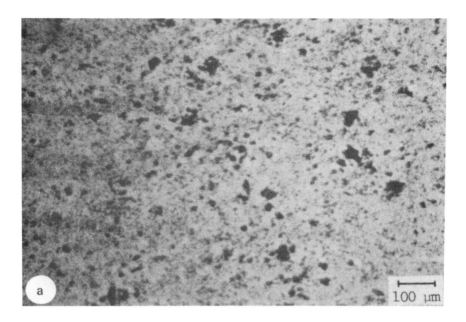

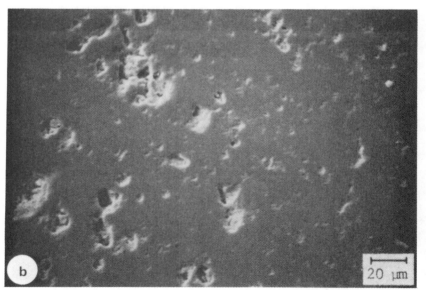

Figure A11. Micrographs of Annawerk Ceranox NR-115H RS-Si_3N_4.

(a) Optical micrograph of polished section
(b) SEM view of polished section

Materials Evaluation 305

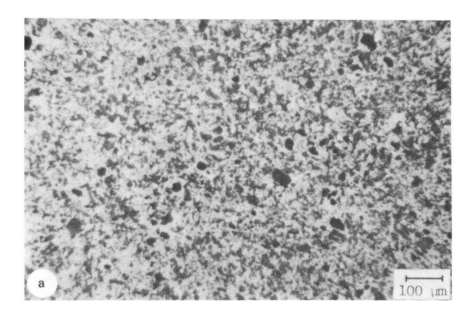

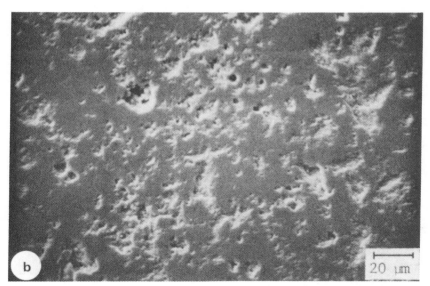

Figure A12. Micrographs of Associated Engineering
 Developments, Ltd., Nitrasil (batch 1) 1978 RS-
 Si_3N_4.
(a) Optical micrograph of polished section
(b) SEM view of polished section

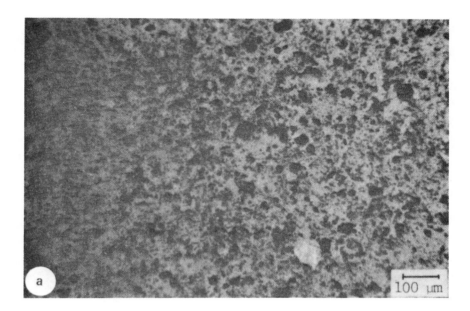

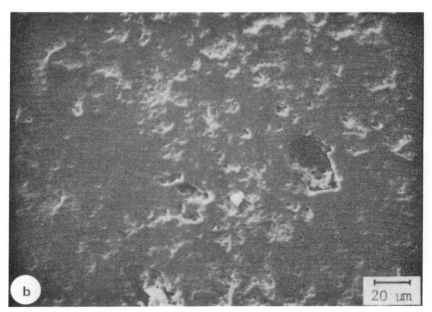

Figure A13. Micrographs of AED, Ltd., Nitrasil (batch 2) 1978 RS-Si_3N_4.

(a) Optical micrograph of polished section
(b) SEM view of polished section

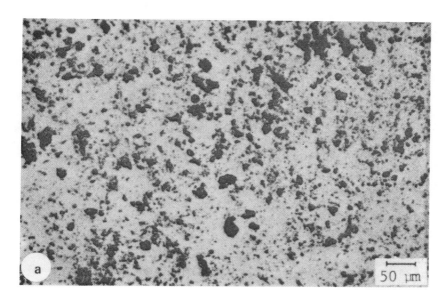
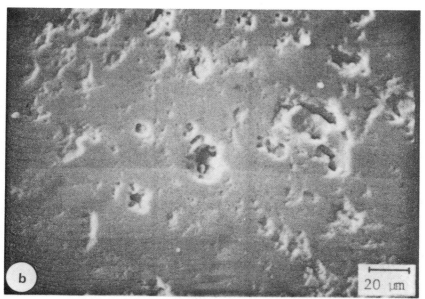

Figure A14. Micrographs of AED, Ltd., Nitrasil
(batch 4) 1978 RS-Si_3N_4.

(a) Optical micrograph of polished section
(gold shadowed)

(b) SEM view of polished section

308 Ceramic Materials for Advanced Heat Engines

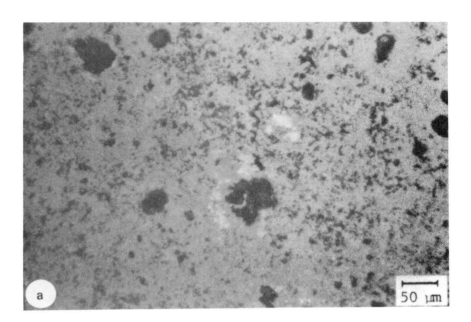

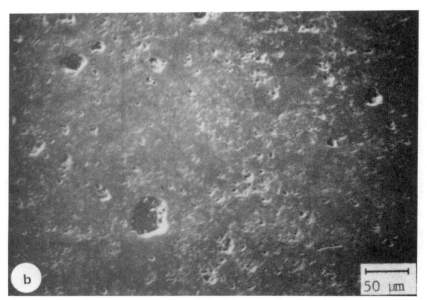

Figure A15. Micrographs of AED, Ltd., Nitrasil (batch 5) 1980 RS-Si_3N_4.

(a) Optical micrograph of polished section
(b) SEM view of polished section

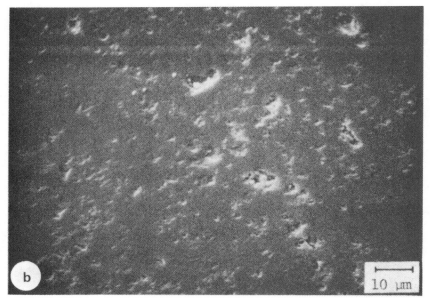

Figure A16. Micrographs of Georgia Tech RS-Si_3N_4.
(a) Optical micrograph of polished section
(b) SEM view of polished section

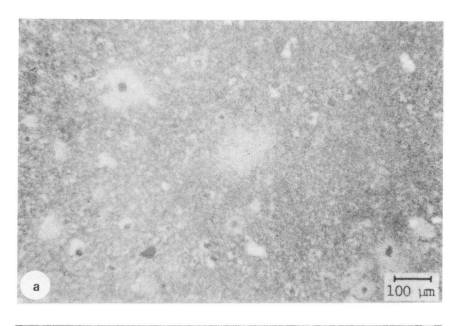
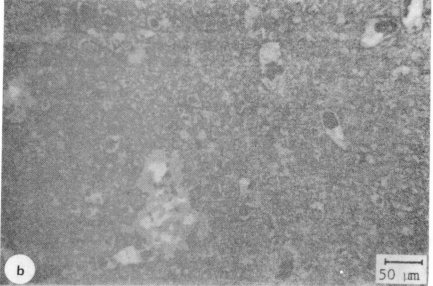

Figure A17. Micrographs of AME RS-Si_3N_4.
(a) Optical micrograph of polished section
(b) Optical micrograph of polished section

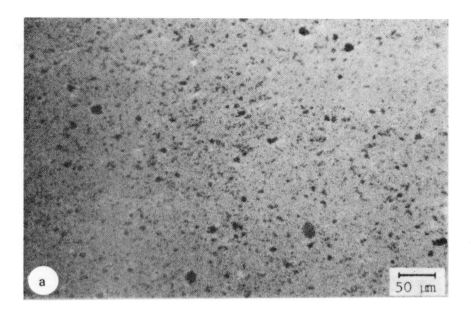

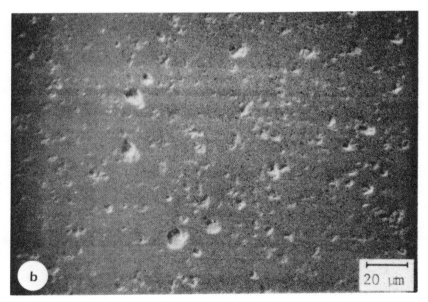

Figure A18. Micrographs of AiResearch RBN-104 RS-Si_3N_4.

(a) Optical micrograph of polished section
(b) SEM view of polished section

Part II

Economic Evaluation

The information in Part II is from *A Structural Ceramic Research Program: A Preliminary Economic Analysis* by L.R. Johnson, A.P.S. Teotia and L.G. Hill of Energy and Environmental Systems Division, Center for Transportation Research, Argonne National Laboratory for the U.S. Department of Energy, March 1983.

Acknowledgments

The authors of this report are pleased to acknowledge the assistance and advice of a number of people. This work was sponsored by the U.S. Department of Energy (DOE), Office of Vehicle and Engine Research and Development. Several people within that office provided useful and prompt reviews including the project manager Anne Marie Zerega, Al Chesnes, and Robert Schulz. Reviews and comments were also provided by Robert Gottschall from U.S. DOE and Vic Tennery, David Greene, and D.J. Bjornstad from Oak Ridge National Laboratory. In addition, Christopher Caton, Mark Booth, and Cary Leahey of Data Resources, Inc. (DRI) provided assistance in formulating the alternative simulations used in the annual version of the DRI U.S. Economy Model.

We also received very valuable insights into ceramic applications from researchers in both industry and at universities. There are too many individuals to mention here but the organizations to which we are indebted include the following:

- AiResearch Casting Company
- Army Materials and Mechanics Research Center
- Caterpillar Tractor Company
- The Carborundum Company
- Coors Porcelain Company
- Cummins Engine Company
- Detroit Diesel Allison
- Ford Motor Company
- The Garrett Company
- General Electric Company R&D Center
- GTE Laboratories
- Illinois Institute of Technology Research Institute
- NASA Lewis Research Center
- Norton Company
- Rockwell International Science Center
- University of Illinois
- University of Washington

Finally, the authors wish to thank LaVerne Schneberger, Shirley Biocic, and Louise Benson for preparing this report and Susan Barr for editing it.

1. Introduction

ABSTRACT

Advanced power systems for vehicles can potentially produce fuel efficiencies that greatly exceed those of today's gasoline and diesel engines. Recent work has focused on adiabatic diesel engines, gas turbine engines, Stirling engines, and electric vehicle batteries. The heat engine technologies, however, are limited by problems involving mechanical strength at high temperatures. Structural ceramics, if they can be reliably mass-produced, can make possible improved vehicle fuel efficiencies through higher-temperature operation and reduced vehicle weight. In this report, the macroeconomic impacts (effects on gross national product, employment, fuel imports, and balance of trade) are modeled for two scenarios, one in which the U.S. dominates the commercialization of ceramics in heat engines throughout the 1990s and the other in which Japan dominates. The positive effects of U.S. dominance were forecast to be substantially greater than the negative effects of foreign dominance due to two assumptions: (1) Japanese ceramic commercialization does not include the truck and stationary engine markets because of a lack of historical presence in these areas and (2) imports of Japanese cars with ceramic engines are legislatively limited to 30% of new car sales. Improved ceramics can also be substituted for superalloys containing strategic materials and thus reduce U.S. dependence on foreign suppliers.

The fuel crises of the 1970s stimulated the development of new technologies and the reexamination of old ones to reduce the U.S. dependence on foreign petroleum supplies. The transportation sector accounts for half of this country's petroleum use and the automobile alone consumes half of the petroleum used for transportation. As a result, both government and industry have made significant efforts to improve existing engines and vehicles and, at the same time, develop advanced concepts. The advanced power systems hold the promise of greatly exceeding the fuel efficiencies of today's gasoline and diesel engines, while also reducing emissions. In recent years, research has focused on the following systems: the adiabatic (uncooled) diesel, the gas turbine, the Stirling engine, and the electric vehicle. In addition, research has been conducted on alternative fuels.

The Office of Vehicle and Engine Research and Development (OVERD) within the U.S. Department of Energy (DOE) has been a major source of research funding for both engine and vehicle development and alternative fuels research. The OVERD emphasizes high-risk, long-range research and technology rather than demonstration and commercialization. One of the projects developed with OVERD support is the advanced gas turbine engine. High-temperature, structural ceramic materials can be used in this engine to allow operation at higher temperatures with associated improved fuel efficiencies.

In the past, OVERD has contracted with engine manufacturers for the engine development. The ceramic fabricators were, in turn, subcontractors who built component parts to meet the specifications of engine manufacturers. This approach, in the view of some industry participants, has generated considerable progress because of the close interaction between the ceramic suppliers and designers; it, however, due to time constraints and funding limitations, has not met the broader need for the systematic improvement of structural ceramic properties and processing techniques. An approach focused instead on the base technology could serve the requirements for several heat engine and other high-temperature applications and be reflected in many related ceramic uses. A research plan[1] incorporating this approach is being developed jointly by Oak Ridge National Laboratory (ORNL) and the National Aeronautics and Space Administration (NASA).

The purpose of this proposed ceramic technology program is to improve both the material properties and production processes in order to produce reliable, cost-effective ceramics to use as high-temperature, structural parts in heat engines, specifically the gas turbine, the uncooled diesel, and the Stirling. Though not a specific objective of the program, it is recognized that the improved ceramics produced through this program can be used for other purposes such as sensors, electronic devices, optical parts, and insulation. Thus, one of the principal advantages of a base technology ceramic research program is that a wide array of products are likely to result from it.

Although research is generally characterized as either basic or applied, the proposed program is described as a base technology research program to indicate a linkage between the basic and applied areas because definitions of these concepts vary widely among researchers. In this report, basic research is defined as research that is concerned with developing an understanding of physical phenomena. For structural ceramic materials, these phenomena include such things as phase equilibria, phase structure, diffusion at high temperatures, effects of stress on microstructure, chemical reactions for synthesizing ceramic powders, surface chemistry, and structure of powder particles. The experimental tools necessary to conduct basic research are generally created or modified as required. Examples of such tools for basic ceramic research include ultrahigh resolution analytical electron microscopes and high resolution ion microprobes.

Base technology research in structural ceramics, in contrast, spans a relatively broad portion of the research and development (R&D) spectrum, from activities that are typically regarded as basic in nature to activities that enhance the understanding of the material applications. In the latter case, data can be used to indicate whether a particular structural ceramic material can be reproducibly fabricated with highly controlled, desirable properties, such as fracture strength, fracture toughness, and erosion resistance at high temperatures. Thus, base technology research provides an important part of the information necessary for industrial decisions concerning future investments in commercial production of the materials.

In the process of developing the proposed program plan, technological needs were assessed; this process included presentations to and discussions with many companies and institutions currently involved in structural ceramics research. The plan has consequently developed in an interative manner as different viewpoints have been incorporated. The ceramic community generally agrees that the research tasks identified conceptually before the assessment was started include all those necessary to achieve the goal of producing reliable, structural ceramics for heat engine applications. These tasks address the following major program elements:

1. Materials and Processing Element

    - Powder synthesis and characterization

    - Fiber synthesis and characterization

    - Processing and characterization of green-state ceramics

    - Densification and characterization of green-state ceramics

    - Characterization of structural properties of dense ceramics

    - Characterization of mechanical and physical properties of dense ceramics

    - Ceramic coating processing and characterization

    - Ceramic-to-ceramic and ceramic-to-metal joining

2. Design Methodology Element

   - Determination of boundary conditions
   - Three-dimensional modeling of components
   - Materials modeling
   - Interface modeling
   - Crack initiation and propagation modeling
   - Alternative statistical approach descriptions
   - New design concept or approach development

3. Data-Base/Life-Prediction Element

   - Generation of statistical property data to improve reliability
   - Generation of design data for component analysis
   - High-resolution nondestructive evaluation techniques

The economic analysis presented here provides a preliminary estimate of the economic effects of a base technology research program in structural ceramics. Inherent in the analysis is the assumption that the proposed research will lead to solutions for known technological problems. Support for this assumption is the concurrence of the ceramic industry and engine manufacturers with the need for the major elements of the proposed research plan.

This economic analysis addresses both the direct and indirect effects that the accelerated commercialization of structural ceramics could be expected to have. The need for federal research support is evaluated and perceptions within the ceramic industry of the problems and potentials that must be addressed are also discussed.

2. Economic Impacts of Structural Ceramic Applications

2.1 METHODOLOGY USED TO ASSESS IMPACTS

The economic impacts of improved high-temperature, structural ceramic materials could be significant as these materials could potentially replace materials currently used in heat engines, heat exchangers, wear parts, and other related applications. The impacts can be put in two categories. First, use of improved ceramic parts could produce "direct benefits" such as productivity and performance improvements. In several applications (e.g., wear parts) the ceramic could increase the life-span of the parts, whereas in others (e.g., heat engines) the improved operation resulting from ceramic use could result in significant fuel savings. Second, an early commercialization by U.S. companies of new parts/products that include advanced ceramic materials could have significant "indirect benefits" such as reductions in the imports of automobiles, increased exports of heat engines, and more jobs.

Table 1 shows the leading U.S. exporting companies and their product areas. Ten of the 13 companies could directly benefit from the commercialization of reliable structural ceramics in heat engines, indicating significant

Table 1   Major U.S. Exporters in 1981

Rank	Company	Products
1	Boeing	Aircraft[a]
2	General Motors	Motor Vehicles & Parts,[a] Locomotives
3	General Electric	Generating Equip.,[a] Aircraft Engines[a]
4	Ford Motor	Motor Vehicles and Parts[a]
5	Caterpillar	Construction Equipment,[a] Engines[a]
6	McDonnell Douglas	Aircraft,[a] Space Systems, Missiles[a]
7	E.I. DuPont	Chemicals, Fibers, Polymer Products
8	United Technologies	Aircraft Engines,[a] Helicopters
9	International Business Machines	Information Systems, Equipment Parts
10	Eastman Kodak	Photographic Equipment and Supplies
11	Westinghouse Electric	Generating Equipment,[a] Defense Systems[a]
12	Signal Companies	Truck, Engines,[a] Chemicals
13	Raytheon	Electronic Equipment, Aircraft[a]

[a] Potential product applications for high-temperature structural ceramic components.

Source: *Fortune*, p. 68 (Aug. 9, 1982).

potential for positively affecting the U.S. balance of trade. On the other hand, early commercialization by foreign companies, at the expense of domestic manufacturers, could have serious detrimental effects for the U.S. In order to measure the direct and indirect benefits of new ceramic technology, a two-step approach was used.

In the first step, the market penetration of ceramic applications was estimated. For each level of penetration of ceramic parts in various end-use markets, the direct benefits, such as fuel savings, were derived. Argonne National Laboratory (ANL) developed the Ceramic Market Model to achieve this objective. Because of the limited time, resources, and/or data, only the heat engine applications are highlighted in this preliminary analysis. The market penetration methodology in general is based on the earlier Argonne work estimating market penetration of (1) new energy-efficient electric motors,[2] and (2) new cogeneration energy systems.[3]

In the second step, the indirect benefits of improved ceramics technology were estimated by using the Data Resources, Inc. (DRI), long-run annual model of the U.S. economy.[4,5] Variables were changed in the model to accommodate different levels of investment, fuel savings, and automobile imports resulting from two different scenarios for the development of advanced structural ceramics. Forecasts were thus obtained for the U.S. economy. By comparing these forecasts with forecasts made using the Base Case Scenario "TRENDLONG2007BANN" of DRI,[6] the indirect benefits were measured for the development of ceramics. The national impact of the commercialization of advanced ceramic materials was measured by examining changes in such key macroeconomic indicators as gross national product (GNP), total employment, total fuel demand, petroleum imports, and total imports.

2.2 POTENTIAL USES FOR HIGH-TEMPERATURE STRUCTURAL CERAMICS

A base technology research program in ceramic technology for high-temperature, structural applications will affect the use of ceramics for many applications beyond heat engines. The potential uses for structural ceramics are grouped below by time frame: near-term, intermediate-term, and long-term. These groupings are approximate; the level of research activities could significantly change the time by which highly reliable ceramic parts could be commercially available.

1. Near-Term (within 5 yr; includes existing uses)

    - Wear Parts – Seals (high-performance and scratch-resistent), bearings (ball and roller), valves, nozzles, liners, pads, gates, and slides

    - Cutting Tools – Tips

- Abrasives

- Corrosion Resistant Parts – Piping, valves, seals, gasifier components (coal), and coatings

- Heat Engine Applications – Diesel wear parts (pushrods, tappets, seals), turbine thermal-barrier coatings, and shrouds that can be abraded

- High Stiffness-to-Weight Structural Parts – Fiber reinforced spars, tiles, and castings

2. Intermediate-Term (by the year 2000)

    - More and Improved Near-Term Products

    - Heat Exchanger Components – for use in metal-melting furnaces, glass-melting furnaces, coal-burning furnaces, and incinerators

    - New Refractories

    - Heat Engine Applications – Diesel combustion zone (cylinders, piston caps), adiabatic diesel, turbocharger rotors, turbine static parts (combustors, shrouds), turbine rotating parts (missiles), and sensors and probes

    - Gun Barrel Liners

    - Nonengine Missile Parts – Radomes, domes, and control valves

    - Corrosion Resistant Parts – Synfuel applications

    - Energy Storage – Electrolytes

3. Long-Term (beyond the year 2000)

    - More and Improved Intermediate-Term Products

    - Heat Engine Applications – Turbine engine rotating parts (rotors, regenerators, military aircraft), Stirling engine parts, Wankel engine parts, and aircraft auxilary power units

    - Solar Concentrator Targets

- Magnetohydrodynamic System Components

- Fusion Reactor Components – First wall blanket

As numerous ceramic materials will be examined during the development of heat engine components, the eventual materials to be commercialized are not yet known. Therefore, these groups certainly do not contain all of the potential uses of improved ceramics. In addition, the proposed research program is likely to lead to improvements in other technical ceramics, including those for electrical/electronic, optical, and low-thermal-conduction uses.

2.3 MARKET PENETRATION FOR SELECTED APPLICATIONS

2.3.1 Selection of the Applications for Analysis

Due to the lack of good market data bases for the many applications that are suitable for structural ceramics, this analysis will focus on a very few applications. This conservative approach could, therefore, be expected to illustrate a lower level of benefits. Ceramics will penetrate these markets because their use will (1) allow higher operating temperatures (leading to improved fuel efficiencies in heat engines), (2) provide lighter weight, and (3) produce better corrosion resistance than will the use of conventional materials. With mass production, the cost of these ceramics is more likely to become competitive with conventional materials. The following uses were selected for this study.

Heat Engine Ceramic Components

This category includes ceramic parts for heat engine applications, especially the combustion zone of the engine, heat exchangers, bearings, and seals. Specifically included for piston engines are pistons, piston rings, valves, cylinder liners, and cylinder-head inserts. The major market for these components is in the transportation and industrial sectors for gasoline and diesel engines. Ceramics are also expected to have significant potential for the diesel and gas turbine engines. For convenience sake, in the analysis it was assumed that the dollar values of the ceramic gas turbine components and ceramic diesel engine components would be the same.

Ceramic Bearings

Bearings are used in a wide variety of applications. However, it is likely that only a portion of the roller (both tapered and nontapered) and mounted bearings used in high-temperature applications would be replaced with ceramic bearings. Ball bearings are much less expensive than ceramic bearings and are not likely to be replaced.

Turbochargers

Turbocharger parts, including the rotor, are expected to have a large market potential. Use of ceramic parts could significantly improve engine performance and fuel economy. Automakers could reduce the number of different engine lines they have by using turbochargers to meet varying engine requirements.

Gas Turbines

Gas turbines are used in a wide variety of applications. This analysis focused on stationary and portable power generation units. Ceramics could be used for the stationary parts and eventually for the rotating parts of these engines. Aircraft gas turbines were excluded from this analysis as ceramic component market penetration is considered by most industry experts to be beyond the year 2000.

2.3.2 Market Potential for Structural Ceramic Applications

The market potential was obtained by estimating the 1981 sales of the conventional-material parts that could be replaced by ceramic parts. Most of the shipment value data for the conventional parts was obtained from the 1977 Census of Manufacturers, issued in July 1980.* The sales were projected to the year 2005 using the growth rates of the industries from the DRI TRENDLONG-2007B Scenario. Table 2 provides the ceramic market potential for the years 1981, 1990, and 2000 by specific applications. The total market potential is estimated to be $4.3 billion in 1981. This market is estimated to increase to $7.2 billion (in 1981 $) by 1990 and to $10.9 billion by 2000. This market growth corresponds to an average annual rate of 6.0% during the decade 1981-1990 and 4.2% for 1990-2000.

2.3.3 Estimated Market Penetration of Structural Ceramic Applications

To estimate the market penetration of any new technology, the competitive aspects of the technology must be considered (i.e., the new technology has to compete economically if it expects to get a share of the market from the other technologies). However, for this analysis, the relative cost data were not available. Therefore, survey analysis methodology was used to estimate the market penetration of new ceramic applications. Industry sources

---

*This is the latest comprehensive canvass of the U.S. industrial and business activities. The data provides the total value of the shipments of products classified as primary to an industry and products that were shipped by other manufacturing firms (secondary to the industry) regardless of their classification (e.g., repair, scrap, etc.).

Table 2 Summary of Market Potential for Selected Ceramic Parts

Applications	Market Potential (10^6 1981 $)			
	1977	1981	1990	2000
Diesel/gas turbine components for highway vehicles				
Piston, piston rings, and valves	595	709	1,199	1,807
Cylinder liners and cylinder head inserts	653	778	1,316	1,984
Bearings				
Taper roller, nontaper roller, mounted bearings, rollers, and spare parts	2,099	2,502	4,232	6,379
Turbochargers	101	120	203	306
Gas turbines, except aircraft	148	176	298	449
Total for all applications	3,596	4,285	7,248	10,925

provided their expectations of the market penetration of various applications by year 2000 under various conditions. Most industry experts believed that only a very small fraction of the market potential could be captured without significant government assistance for the basic research necessary to overcome the technological barriers currently associated with the ceramic materials. Section 4 of this report details this information.

Table 3 shows a range of market penetration estimates. The first ("low-penetration") estimate assumes little or no government funding for ceramic research conducted by public and private organizations. Under this scenario, only 5% of the total market will be penetrated for various applications. However, industry sources (as discussed in Sec. 4) believe that a determined effort spurred by government funding for ceramic research could lead to commercialization by the late 1980s and 1990s. Under this "high-penetration" scenario, the penetration could be 20-30% for most market applications by 2000 and up to 75% for the turbocharger application.

The relationship of market penetration to market potential for these high and low cases is shown in Fig. 1. The ceramic market penetration was assumed to begin in 1985 for diesel engines larger than 200 hp (truck and

Table 3 A Range of Market Penetration for Ceramic Components by the Year 2000

Applications	Market Potential (10^6 1981 $)	Market Penetration (%)		Market Penetration (10^6 1981 $)	
		Low	High	Low	High
Diesel/gas turbine components					
Piston, piston rings, valves[a]	1,807	0.05	0.20	90	361
Cylinder liners, head insert[a]	1,984	0.05	0.20	99	397
Bearings	6,379	0.05	0.30	319	1,913
Turbochargers	306	0.05	0.75	15	230
Gas turbines, except aircraft	449	0.05	0.30	22	135
Total	10,925			545	3,036

[a]Includes gasoline and diesel engine parts for all applications.

stationary), bearings, and turbochargers. Subsequently, with the demonstration of the technology in the marketplace, the ceramics were assumed to begin by 1990 to penetrate the more complex markets of small diesel engines (<200 hp), adiabatic diesel engines, automobile diesel engines/gas turbines, and stationary gas turbines. The ceramics market penetration shares were assumed to follow well-known, new-technology, S-shape patterns.

Under the "low" scenario, the total market penetration is estimated to approach $545 million (1981 $). For the "high" scenario, this penetration would be over $3 billion (1981 $).

2.4 ECONOMIC EFFECTS OF ALTERNATIVE SCENARIOS FOR CERAMIC COMMERCIALIZATION

With a consensus of those in the ceramic industry and material science field that it is only a matter of time before the widespread commercialization of structural ceramics, it becomes increasingly important to consider the effects of who captures the market first. Significantly different benefits could be expected depending upon whether the U.S. or foreign companies reach the marketplace initially. Not only are there direct benefits to individual firms that sell (and profit) from the ceramic components, but there are

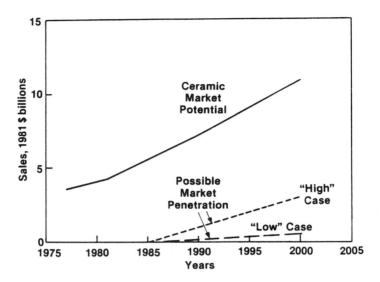

Fig. 1 Market Penetration for Ceramic Applications under Alternative Cases

aggregate impacts at the national level. Changes in GNP, employment, and balance-of-trade accounts are quantifiable indicators of the influence of domestic vs. foreign dominance in any product area.

Economic impacts can be expected from the development of more reliable ceramic parts. For example, the life of wear parts such as bearings could be increased significantly. In other applications, use of advanced ceramics would permit higher performance of the parts or assemblies. For heat engine applications, significant fuel savings could be achieved through the use of ceramic parts, due to lighter weight, reduced friction, and, for piston engines, elimination of a cooling system. These impacts can be categorized as "direct benefits." Indirect economic gains could also result from this new ceramic technology. If the U.S. industry is able to exploit these materials first, improvement in the state of the domestic industries and substantial gains in the U.S. economy could result. One beneficiary could be the U.S. automobile industry, which could gain a technological advantage against the Japanese automobile companies by taking the lead in the commercialization of ceramic engine components. Conversely, foreign dominance in the commercialization of structural ceramics would have a direct negative impact on the domestic auto industry.

2.4.1 Assumptions

Two different scenarios were constructed for this study, representing two divergent views for the commercialization of structural ceramics. These scenarios bracket a range of potential ceramic developments for the U.S. and Japan. This parametric analysis is not intended to assume either outcome as probable but only to define the limits of likely economic effects. This approach was used because of the infinite number of assumptions that could be postulated about each country's R&D funding, research success, market strategy, and eventual market penetration and share. In addition, a "Base Case" scenario was used for comparison. This Base Case scenario assumes no particular efforts with respect to the development and marketing of structural ceramics. In the first scenario or "U.S. Dominant Case," there is considerable industry/government cooperation and, in effect, a national R&D effort for structural ceramics. Excluding military applications, ceramic components are assumed to penetrate the market for stationary and heavy truck engines beginning in 1985. Subsequently, the ceramic diesel engine (uncooled) and/or the gas turbine will begin to penetrate the automotive and light truck markets by 1990. The adiabatic diesel (uncooled, but operating at higher temperatures with little heat transfer) is also assumed to be available in limited supply about the same time. The assumptions for this scenario are identical to those for the heat engine applications under the high-penetration scenario discussed in Sec. 2.3.

This optimistic scenario assumes a reversal of the present direction of the Japanese research in ceramics. Such a change could be prompted by their recent economic difficulties and budget constraints. (Although the current economic problems in Japan are moderate by the standards of major industrial countries, they are very serious by Japanese standards, with record unemployment, increasing inflation, and weak currency. A $40 billion budget deficit is expected this year and the Finance Ministry is attempting to cut research funding for committed projects as one of several alternatives to controlling the government deficit.) As a result of the U.S. lead in ceramic research during the 1970s and Japan's limited national funds for research, they are assumed to focus on other areas. The U.S. success during the 1990s is assumed to cause them to renew their efforts so that by the end of the decade ceramic engine components will be available for the cars they export to the U.S. The decline in cars imported to the U.S. is then assumed to be stabilized.

The other scenario is a pessimistic view of ceramic developments. The "Foreign Dominance Case" assumes no concerted national effort in the U.S., while other countries, especially Japan, continue with substantial government-funded research programs in ceramics. In this scenario, U.S. industry, while recognizing the potential for structural ceramics, considers the market too risky because of the strong government support. Because Japan has traditionally focused on engines with low horsepower requirements, their ceramic commercialization is assumed to begin with automobile engine components in 1985. The import share of automobiles will increase at the expense of domestic car makers. However, the market share for imported cars is not expected to exceed 30%, at which point trade restrictions or licensing agreements would be anticipated. Foreign (i.e., Japanese) domination of structural ceramics in the heat engine market is assumed to be confined to the automotive and light truck applications. Japanese penetration in the U.S. markets for heavy-duty trucks or stationary diesel/gas turbine engines is not expected without the presence of a North American manufacturer. Figure 2 provides an historical, as well as comparative, view of the scenarios for U.S. auto imports.

Because of the size of the automobile market (compared to the truck or stationary engine markets), the most significant sales potential and consequent national economic effects can be expected in this area. Because different assumptions of market penetration were made for each scenario, the ultimate economic impacts are more different than might be expected given the assumption of similar market penetrations but with different countries leading. Table 4 gives the new car sales for each scenario. In the U.S. Dominant Case, the cars with ceramic engines achieve a market share of 20% by the year 2000. Domestic sales at this level would reduce the share of imported cars to 15%. On the other hand, when ceramic engine car sales reach 500,000 annually in the Foreign Dominance Case, imported cars would then account for about 28% of U.S. sales -- a level that is assumed to result in politically imposed

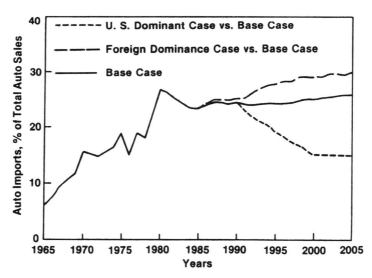

Fig. 2 U.S. Automobile Imports

Table 4 Market Penetration of Automotive Ceramic Component Engines for Alternative Scenarios

Year	Base Case		U.S. Dominant		Foreign Dominance	
	Total Sales (10^6)	Imports (%)	Ceramic Sales (10^3)	Imports (%)	Ceramic Sales (10^3)	Imports (%)
1985	10.9	23.4	0	23.4	10	23.4
1990	11.9	24.5	125	24.4	100	25.2
1995	12.6	24.2	1270	19.0	500	27.8
2000	12.7	25.2	2540	15.0	500	29.1
2005	13.0	25.9	3580	15.0	500	30.0

restrictions. Keeping the number of imports constant at that level through the 1990s would prevent the import share from exceeding 30%.

These two scenarios bracket the range of reasonable successes for structural ceramics in the heat engine market. The results of simulating these two cases in the DRI model are presented below and compared to the base case (i.e., no particular effort to market structural ceramics). Specific information concerning scenario changes in the model is provided in Appendix A.

2.4.2 Macroeconomic Effects of Ceramic Penetration in the Heat Engine Market

Base Case Macroeconomic Scenario

The scenario used for comparison purposes is the latest DRI long-term outlook of the U.S. economy, TRENDLONG 2007 BANN.[6] The highlights of this scenario are summarized in Table 5.

In general, under this scenario, the economy exhibits mild variations in growth and remains lower than its historical trend path. The inflation level moderates slowly from the levels of the last few years to a lower level, but remains higher than the average level of the past 25 years. Unemployment is also forecast to remain at a higher level for the next 25 years than it has for the past quarter of a century.

Table 5 Base Case Economic Indicators

	%, Growth	
Economic Indicator	Actual, 1956-1981	Forecast, 1982-2007
Real Gross National Product	3.3	2.5
Inflation, GNP deflator	4.6	6.1
Unemployment	5.6	6.9
Industrial Production	3.7	3.0
Labor Force	2.0	1.1
Fuel Import Cost as % of GNP	1.0	2.8

Macroeconomic Effects - U.S. Dominant vs. Foreign Dominance Cases

The major national effects of a successful U.S. structural ceramics research program and of a Japanese lead in ceramic commercialization were determined by solving the DRI model, using the general assumptions described earlier and the specific assumptions detailed in Appendix A.

Real Gross National Product. Under the U.S. Dominant Case as compared to the Base Case, the economy performs better, as this policy is expansionary. As shown in Fig. 3, the real GNP is higher by $2.5 billion (1981 $) in 1990, $15.2 billion in 1995, and $28.2 billion higher in 2000. The gap between the U.S. Dominant Case and Base Case narrows to $16.6 billion by 2005 because of cyclical effects. The high growth in the economy in earlier periods is inflationary (GNP deflator), which dampens the economy in the later periods. Between 1985 and 2005, the cumulative real GNP gain (U.S. Dominant Case over Base Case) is $278.9 billion (1981 $). As Fig. 3 shows, the contribution of ceramic technology to the real GNP is enormous, even when the future gains are discounted by an appropriate real discount rate.

Under the Foreign Dominance scenario, real GNP follows a lower trajectory than that under the DRI Base Case scenario. The real GNP is down by $0.2 billion (1981 $) in 1985, $2.2 billion in 1990, $11.1 billion in 1995, and $6.5 billion by 2000 as shown in Fig. 3. In the following years, the long-run equilibrium forces in the economy narrow the gap to $2.7 billion by 2005. The imports of additional ceramic engine automobiles at the expense of

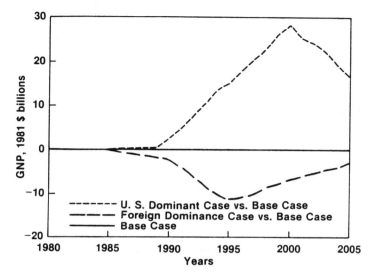

Fig. 3 Gross National Product under U.S. Dominant and Foreign Dominance Scenarios Compared to Base Case Scenario

domestic car makers shift the production to the foreign countries. Import spending is thus higher, which reduces GNP. Between 1985 and 2005, the cumulative real GNP loss (Foreign Dominance Case over Base Case) is estimated to be $110.7 billion (1981 $).

Total U.S. Employment. Because of the high labor requirements of the automobile industry, employment increases under the U.S. Dominant Case as compared to the Base Case (Fig. 4). The total employment rises by 25,000 in 1990, 174,000 in 1995, and 250,000 in 2000. The national unemployment rate drops from 6.5% under the Base Case to 6.3% under the U.S. Dominant Case in year 2000.

The lower level of economic activity in the Foreign Dominance Case results in lower employment levels. Compared to the Base Case scenario, the total U.S. employment declines by 2000 jobs in 1985, 13,000 jobs in 1990, and 106,000 jobs in 1995. As imports of ceramic engine automobiles stabilize at a level of 500,000 cars/yr beyond 1995, the employment loss bottoms out in 1995 (Fig. 4). By the year 2000, the economy tends to move towards long-run equilibrium. The maximum loss of 106,000 jobs in 1995 under the Foreign Dominance Case compares with a peak gain of 250,000 jobs in 2000 under the U.S. Dominant Case.

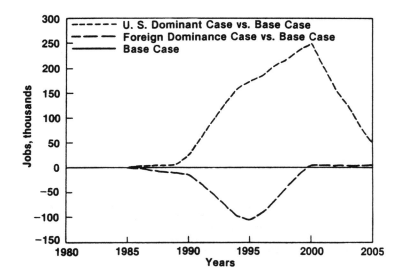

Fig. 4 U.S. Employment under U.S. Dominant and Foreign Dominance Scenarios Compared to Base Case Scenario

<u>Total U.S. Imports Savings</u>. Significant savings in total imports are realized under the U.S. Dominant Case compared to the Base Case (Fig. 5). These imports savings are approximately 0.7, 5.5, 14.4, and 27.7 billion (1981 $) in 1990, 1995, 2000, and 2005, respectively. The impacts of the reduction in imported cars peaks in 2000, while fuel imports continue to fall even by 2005 as stock of more-efficient cars and trucks with ceramic engines and parts keep rising.

Under the Foreign Dominance Case, the imports implications are highly negative. Because of the influx of the ceramic engine automobiles, the U.S. expenditures for imports remain greater than under the Base Case scenario as shown in Fig. 5. The total imports savings shown are net saving in the economy. The net imports are greater by 0.8, 3.3, 4.5 and 5.5 billion (1981 $) in 1990, 1995, 2000, and 2005, respectively.

<u>Total Fuel Savings in the Economy</u>. Between 1985 and 1990, the fuel savings in the U.S. Dominant Case are negligible as the fuel efficiency of ceramic engines is offset by the greater demand for fuel in the economy, which is on a moderately expansionary path as a result of higher investment of ceramic-related expenditure, (e.g., new assembly lines). However, as ceramic

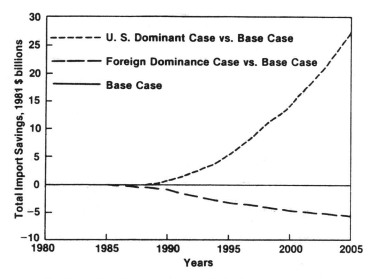

Fig. 5 Total Import Savings under U.S. Dominant and Foreign Dominance Scenarios Compared to Base Case Scenario

engine stocks build up, the fuel savings continue to increase. The total fuel savings approach 0.1 quad by 1995, 0.25 quad by 2000, and 1.1 quad by 2005.

The fuel savings in the economy for the Foreign Dominance Case are negligible in comparison to the total fuel use, due primarily to the assumed restrictions on imported car sales after a market share of nearly 30% is achieved. The direct energy saved by the fuel efficient ceramic engine cars increases from 0 in 1985 to 0.006, 0.030, 0.055, and 0.058 quad in 1990, 1995, 2000, and 2005, respectively. However, the total fuel demand in the economy will depend on the state of the economy. As the real GNP (Fig. 3) under the Foreign Dominance Case compared to Base Case declines during 1985-1995 period, the U.S. annual fuel savings reach 0.1 quad by 1990 and 0.3 quad by 1995. However, as the penetration of imported ceramic engine cars peaks in 1995, the real GNP gap between the Foreign Dominance Case and Base Case narrows in the period 1995-2005, leading to the relatively greater demand for energy in the economy. Consequently, the U.S. fuel savings shrink to 0.20 quad by 2000 and 0.10 quad by 2005.

Import Fuel-Bill Savings. The substantial reduction in the fuel demand (diesel and gasoline) associated with a U.S. ceramic program is projected to have a very favorable impact on the fuel import bill. As shown in Fig. 6, the import fuel bill will be lower by $10.2 billion (1981 $) in 2005.

In the Foreign Dominance Case, fuel imports savings grow from 0 in 1985 to $1.8 billion (1981 $) in 1995 because of (1) energy savings in fuel efficient ceramic engine automobiles, and (2) a generally weaker economy during this period. However, as the imports of ceramic engine cars peak in 1995 and the economy tends towards the Base Case long-run equilibrium path, the fuel imports savings shrink in the period 1995-2005. In 2005, the fuel imports savings stand at $0.6 billion (1981 $).

Conclusions. The analysis indicates that U.S. commercialization of structural ceramic materials in heat engines has significant rewards in terms of increasing the real GNP (by $279 billion in 1981 $ during the first 20 yr of market penetration). In addition, sizeable savings are realized in total imports (by $28 billion in 1981 $ in 2005). This saving is split, more or less, between reduced imports of automobiles and crude oil. Private investment in ceramics technology of approximatly $6 billion (1981 $) is, however, a high-risk, high-reward situation. Whether ceramics can be commercialized between 1985 and 1990 may depend on the efforts and commitments of both

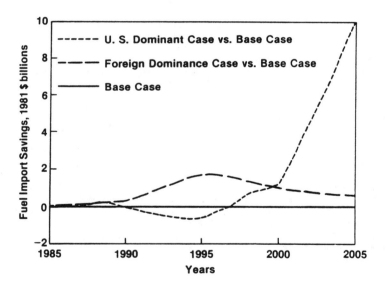

Fig. 6 Fuel Import Savings under U.S. Dominant and Foreign Dominance Scenarios Compared to Base Case Scenario

private industry and the government in the research, as will be further discussed in Secs. 3 and 4 of this report.

Conversely, giving the advantage to foreign competitors, particularly to Japan, could have serious impacts on the economy. The U.S. economic output could decline as much as $111.0 billion (1981 $) in the period of 1985-2005. This approach also implies that the U.S. could lose as many as 106,000 jobs in its worst annual impact. This loss is in sharp contrast to the economic gains shown under the U.S. Dominant Case, which adds as many as 250,000 jobs to total employment in its most favorable annual impact.

2.5 STRATEGIC MATERIALS IMPACTS

In a recent report, the U.S. General Accounting Office (GAO) stated: "The pursuit of our (U.S.) most important national goals dictates that we be concerned about materials, and that we make an effort to develop an enlightened materials policy. Materials availability and prices will affect our success in trying to reach a full-employment economy. Critical materials are essential to our goal of balanced economic growth. They can aid in our effort to reduce inflation. They affect our balance of trade. They will be crucial to our relations with emerging nations all over the world, and they have profound impacts on our environment. Ultimately, they will determine our success in attaining sustainable levels of production and consumption."[8] Among the materials mentioned as critical are the following: beryllium, chromium, nickel, palladium, platinum, titanium, and tungsten. Two basic approaches can be used to critical materials problems: (1) stockpiling and (2) finding suitable substitutes. The proposed base technology research program in ceramics, if successful, could provide an adequate substitute for these materials in many applications.

Materials that may be affected by new developments in ceramic research are listed in Table 6. The following discussion provides basic information about these materials, a rationale for their classification as critical, and an assessment of the possible impact of the ceramic technology program on each material.

2.5.1 Beryllium

Beryllium is a high-strength, lightweight metal with excellent anticorrosion characteristics. It is suitable for a large number of industrial applications; however, high cost and processing problems have limited its use to the most demanding applications. The U.S. does stockpile beryllium metalgroup ore and metals as indicated in Table 7. In the United States, the primary producers of beryllium are Brush Wellman, Inc. (BWI), and Kawecki Berylco, Inc. (KBI), and production figures are withheld from Bureau of Mines (BOM) reports.[9] Table 8 shows that the U.S. reserves base figures are roughly 7% of the total world reserves. Thus, the U.S. is likely to continue

Table 6 Strategic Materials for which Ceramics Could Substitute

Material	Symbol	Application where Ceramics May Replace the Material	Technological Barriers for Ceramics
Tungsten	W	Wear parts (liners, pads, nozzles, bearings, gates, slides, valves, seals).	Demonstration, scale up, cost
		Cutting tools and abrasives	Achieve hardness potential of Si_3N_4
Cobalt	Co	Diesel combustion parts	Demonstration, cost
		Turbocharger rotors	N/A
		Heat recovery systems	N/A
Nickel	Ni	Diesel combustion parts	Demonstration, properties; maintenance of current creep and oxidation levels, cost
		Turbocharger rotors	N/A
		Heat recovery systems	Fabrication and performance innovations, demonstration, cost
		Chemical ware (anti-corrosion piping, valves, seals, etc., gasifier components, coatings)	N/A
Chromium	Cr	Heat engines	Demonstration, properties, cost
		Chemical Ware	N/A
		Heat Exchangers	Fabrication, performance, demonstration, cost
Molydbenum	Mo	Heat recovery systems	N/A
Manganese	Mn	Heat engines	Demonstration, properties, cost
Titanium	Ti	Wear parts	Demonstration, scale-up, cost
		Cutting tools and abrasives	Achieve hardness, new composites
		Heat Applications	N/A
Platinum and Palladium	Pt,Pd	Chemical ware	N/A
Beryllium	Be	Structural members (fiber reinforced spars; tiles; castings)	Innovation in composites, organoceramics technology development
Columbium (Niobium)	Nb	High-temperature applications	Fabrication, performance, demonstration
Tantalum	Ta	High-temperature application	Costs, hardness potential
		Wear parts	N/A

N/A - Not Available.

Table 7 Characteristics of Selected Strategic Materials Imports

Mineral	1975-1980 Range of Net Imports (% of apparent consumption)	Stockpiled by U.S. Government	World Resource Adequacy to Year 2000	U.S. High-Grade Resource Adequacy to 2000: Problem	U.S. Vulnerability to Foreign Disruption	Major Import Sources
Beryllium	N/A	Yes	No	Major	No	Brazil, Republic of South Africa
Cobalt	100%	Yes	No	Major	Yes	Zaire, Zambia
Chromium	90%-100%	Yes	No	Major	Yes	South Africa, U.S.S.R., Turkey, Zimbabwe
Manganese	98%	Yes	No	Major	Yes	Gabon, Brazil, South Africa
Molybdenum	0%	No	No	No	No	Net Exporter
Nickel	70%-80%	Yes	No	Major	Maybe	Canada, New Caledonia, Dominican Republic, Australia
Columbium (Niobium)	100%	Yes	No	Major	Yes	Brazil, Thailand, Canada
Platinum and Palladium	90%	Yes	No	Major	Yes	South Africa, U.S.S.R., England
Tantalum	98%	Yes	Maybe	Major	Yes	Thailand, Canada, Australia, Brazil, Zaire
Titanium	N/A	Yes	No	No	No	Canada, Australia
Tungsten	45%-60%	Yes	No	Minor	No	Canada, Bolivia, Peru, Thailand

N/A - Not Available.

Source: Ref. 11.

Table 8 Economic Characteristics of Selected Strategic Materials
(Bureau of Mines Mineral Statistics)

Material	Year and Units of Measurement	Reserve Base Resources World	Reserve Base Resources U.S.	Primary Production World	Primary Production U.S.	Primary Consumption in U.S.	Import (%)	Probable U.S. 2000 Demand	Per Unit Price	2000 Probable Transportation Demand	2000 Probable Aerospace Consumption
Beryllium[a]	1978-short ton	419,000	28,000	98	N/A	70	N/A	180	$98/short ton	N/A	20
Chromium[b]	1980-10³ short ton	650,000	50	2,830	0	365	100	1,100	$121/short ton	200	N/A
Cobalt[c]	1980-10³ lb	3,300,000	0	72,564[d]	0	15,321	100	28,600	$25/lb	N/A	5,200
Columbium (Niobium)	1978-10³ lb	22,000,000	0	23,700	0	5,700	100	N/A	$3.18/lb	N/A	N/A
Manganese[e]	1980-10³ short ton	6,000,000	0	25,000	0	1,190	73	2,130	$1.70/lb	450	N/A
Molybdenum	1980-10³ short ton	17,000,000	7,500,000	215,482[f]	148,516	50,289	net exporter	170,000	$9.70/lb	46,000	N/A
Nickel[g]	1980-10³ short ton	60,000	200	648[f]	15	156	77[f]	580	$3.41/lb	104	55
Palladium[h]	1977-10³ troy oz	420,000	N/A	2,950[i]	5	516	91	1,125	$4.39/oz	130	0
Platinum[h]	1977-10³ troy oz	800,000	1,000	2,707[i]	5	721	91	1,175	$2.14/oz	0	0
Tantalum[j]	1978-10³ lb	144,000	0	2,200	0	1,610	100	2,980	$33/lb	110	N/A
Titanium[k]	1980-10³ short ton	80,000	1,500	1,736[l]	236[l]	535	N/A	1,414	$7.02/lb	N/A	28
Tungsten[m]	1980-10³ lb	3,920,000	238,000	95,000	6,036	20,200	50	51,800	$1.31/short ton	5	N/A

[a]Ref. 9.
[b]Ref. 13.
[c]Ref. 10.
[d]1976 figure.
[e]Ref. 14.
[f]1978 figure.
[g]Ref. 15.
[h]Ref. 16.
[i]1977 figure.
[j]Ref. 17.
[k]Ref. 18.
[l]Rutile only reserves.
[m]Ref. 19.

importing and stockpiling this material. Table 8 also shows that U.S. demand is expected to increase by 250% from 1978 to 2000. The development of ceramics could reduce the use of beryllium in some lightweight and anticorrosive applications.

2.5.2 Cobalt

Cobalt is a little-known strategic metal. Originally used as an obscure coloring additive, it is now considered an essential element in many alloys and an important ingredient in chemical compounds.[10] In most of its alloying applications, cobalt imparts qualities such as heat resistance, high strength, wear resistance, and magnetic properties. Major uses include jet engine parts, catalysts, and pigments and dryers for paints and allied products. The basic U.S. problem regarding cobalt is excessive dependence (98-100%) on foreign sources of supply because of the concentration of the high-grade deposits in only a few areas of the world. Public Law 96-536, Continuing Appropriations for Fiscal Year 1981, contained a $100 million appropriation for the U.S. stockpile acquisitions program. This allocation was used exclusively to purchase cobalt because of its high priority. Table 7 also shows that the U.S. has no major source of high-grade cobalt. Table 8 shows that demand for cobalt is estimated to increase significantly. Thus, the U.S. dependence on cobalt can be expected to continue until significant new sources or substitutes are discovered. One possible cobalt source is nodules that could be mined from the floor of the Pacific Ocean, but many legal and political questions about this source remain unanswered. Another solution could be found in advanced ceramics. As can be seen in Table 6, materials researchers predict that cobalt alloys could be replaced by ceramics in diesel combustion engines, turbocharger rotors, and heat recovery systems if costs and reliability are improved. Table 9 shows superalloys commonly used in the aircraft industry. Many of these alloys contain cobalt. If ceramic components could be used to replace the components built of the superalloys, significant reductions in cobalt usage could occur. Table 8 shows that the cobalt demand by the aerospace industry projected for 2000 is 5,200 lb or 18% of all the cobalt demand.[12] The use of improved ceramic components could cause that demand to decline by 5% or 1300 lb.

2.5.3 Chromium

Chromium is one of the most versatile elements in modern industry. In the metallurgical, chemical, and refractory industries, it is used for such items as heating elements, pigments, leather processing catalysts, and furnace linings.[13] Almost every list of critical materials includes chromium. As shown in Tables 7 and 8, the U.S. imports nearly all of its chromium from potentially unstable suppliers (South Africa, Zimbabwe, Turkey, and Russia), and high-grade domestic resources are insignificant. Furthermore, demand in 2000 is estimated to increase considerably from current levels. Many of the

Table 9 Material Composition in Selected High-Strength, High-Temperature Alloys Used in the Aerospace Industry

Alloy	Material in Alloy (%)					General Use
	Manganese	Chromium	Nickel	Cobalt	Tungsten	
21 Alloy	1.0	29.0	3.75	55.0	0	Good high-temperature and shock resistance. Oxidation resistance to 2100°F.
X-40	1.0	24.5	9.5	10.0	7.0	Maximum high-temperature strength. Oxidation resistance to 2100°F.
Hastelloy	1.0	15.5	41.0	2.5	3.75	Intermediate high-temperature strength. Oxidation to 2100°F. Resistance to thermal shock.
N-155	1.0	20.0	19.0	18.5	2.0	Good strength at intermediate temperature. Oxidation resistance to 2000°F.
309 Mod	1.0	22.0	11.0	0	2.5	Good strength at intermediate temperature with low alloy content.
Inconel X	1.0	16.0	70.0	0	0	Maximum elevated temperature strength properties in heat-treated conditions.

Source: Ref. 12.

superalloys listed in Table 9 contain large amounts of chromium. Because these superalloys are used in many high-temperature applications including military aircraft and missiles, chromium is critical to both the U.S. economy and defense. As Table 5 shows, the use of ceramic components could decrease the need for chromium-containing superalloys in such applications as heat engines, chemical ware products, and heat exchangers, thus lessening U.S. vulnerability.

2.5.4 Manganese

The principal use for manganese is in the production of steel, where it is an essential element. Tables 7 and 8 show that the U.S. has no reserves of high-grade manganese ore. The U.S. does, however, have several large low-grade deposits that are believed to contain manganese suitable for refractory applications.[14] Ceramics can probably be substituted for some manganese-containing superalloys in heat-engines if demonstration and cost problems can be overcome.

2.5.5 Nickel

Without nickel or an effective substitute, the sophisticated industrial complexes that provide our standard of living and military armament would not be possible.[15] The greatest value of nickel is in alloys with other (generally critical) material elements, where it adds strength and corrosion resistance over a wide range of temperatures. Tables 7 and 8 show that the U.S. imports 70-80% of its nickel needs from Australia, Canada, New Caledonia, and the Dominican Republic and has few domestic high-grade reserves. The BOM has estimated the probable U.S. demand for this material at 104,000 short tons in 2000. Table 5 shows that nickel-containing alloys could be replaced by a high-temperature ceramic in diesel engines, turbocharger rotors, heat recovery systems, seals, and anticorrosion parts if cost and composition problems can be overcome. Table 9 shows the percentage of nickel in the alloys most commonly used in the aerospace industry.

2.5.6 Columbium (Niobium)

Columbium is used in the steel and aerospace industries.[20] Basically, this element imparts strength and high-temperature resistance to superalloys and other low alloy materials. Tables 7 and 8 show that the U.S. imports 100% of its columbium needs from Brazil, Thailand, and Canada and has no domestic high-grade reserves. The BOM has not estimated the probable U.S. demand for this material. Table 5 shows that alloys containing columbium could also be replaced with a suitable ceramic in many high-temperature applications. As the U.S. imports 100% of its columbium, any reduction in usage will reduce the need for imports.

2.5.7 Platinum and Palladium

The platinum group consists of six closely related metals, which commonly occur together in nature.[16] Two of these metals are platinum and palladium. These are among the scarcest of metallic elements, and, as a result, their cost is high. Of the U.S. annual supply of platinum-group metals, about 10% is secondary metal, derived mainly from domestic sources, and 90% is primary, or newly mined metal, virtually all of which is imported. Tables 7 and 8 show that U.S. mine production and reserves are small and that the U.S. is heavily dependent on the Republic of South Africa and the U.S.S.R. for supplies of these critically important strategic industrial metals. The outflow of dollars ($275 million in 1977) to buy these metals adversely affects the U.S. balance of trade. Therefore, the need to develop substitutes for this group of metals is critical. Ceramics could replace these metals in chemical ware parts. The chemical industry uses over 25% of all the palladium and 10% of the platinum in the U.S. Thus, some savings may be made in the balance of payments if suitable ceramics become available.

2.5.8 Tantalum

Tantalum is a refractory metal with unique electrical, chemical, and physical properties that dictate its use in such items as electronic components, metal-working machinery, chemical equipment, and nuclear reactor components. Tantalum is a ductile and easily fabricated metal, has a high melting point (2996°C), and is highly resistant to corrosion by all acids except fuming sulfuric acid and acids containing fluorine.[17] It combines readily with other refractory metals (e.g., tungsten and hafnium) to form alloys having high-temperature strength and stability, although they are not generally oxidation resistant. As Table 6 and 7 show, tantalum raw materials are not produced domestically. The U.S. depends upon imports of tantalum-containing concentrates and tin slags for its primary tantalum supply. The imported supplies come chiefly from Canada, Australia, Brazil, Thailand, and Zaire. Suitable ceramics can be expected to replace tantalum in cutting tools and wear parts. At $33/lb (for contained tantalum in concentrate) substantial replacement of this material by suitable ceramics would have a positive effect on the U.S. balance of payments.

2.5.9 Titanium

About 65% of the titanium metal consumed in the U.S. is for aerospace applications, including aircraft and guided missile assemblies, spacecraft, and turbine engines.[18] The remainder is used in the chemical and electrochemical processing industry, power plants, marine and ordinance applications, and steel and other alloy manufacturing. The mineral sources of titanium products are rutile and ilmenite. Concentrates of these minerals are made at a relatively small number of operations throughout the world. As only a small number of firms mine titanium in the U.S., domestic production is not reported. It is known, however, that the U.S. is an importer of the metal, primarily from Canada and Australia. Table 6 shows that suitable ceramics could replace titanium in heat applications, wear parts, chemical and electrochemical processing, and cutting tools. The production of the suitable ceramics would require a research program to achieve hardness of the ceramic, reduce costs of production, and demonstrate that the products could be "scaled up."

2.5.10 Tungsten

The unique high-temperature properties of tungsten suggest significant increased demand in all of the three predominant end uses: carbides, ferroalloys, and metal.[19] Although production of tungsten concentrate has been equal to approximately one-half to two-thirds of the reported U.S. demand, the rate of production will likely stay relatively constant while demand will increase at a faster rate than the industrial economy grows. As a result, the U.S. will become increasingly dependent on imports and sales of current government stockpiles. As shown in Table 6, the tungsten demand could be reduced by the development of a suitable ceramic to replace it in wear parts such as seals and valves.

2.5.11 Summary of Overall Impacts

Solution of U.S. critical materials problems will depend in part on the development of appropriate substitutes for these materials. Empirical study of the extent to which ceramics can replace these materials is necessary. However, the preliminary assessment here suggests that advanced structural ceramics could substantially decrease consumption of many of the materials and improve the U.S. trade balance.

2.6 FOREIGN COMPETITION

One of the principal concerns of the American ceramics industry is their perception that U.S. support for research in structural ceramics is declining at a time when foreign governments are establishing long-range programs. England was in the forefront in technical ceramics research during the last decade; however, both the level of government support and the pace of development have decreased in the last 10 yr. The U.S., West Germany, and Japan are currently the major competitors in an effort to commercialize structural ceramics.

Many observers characterize Japan as the current leader with West Germany not far behind. One indication of recent developments in high technology ceramics is a comparison of commercially available polycrystalline partially stabilized zirconium oxide (PSZ), a ceramic with potential applications in some heat engines. As shown in Table 10, Australia, Japan, and West

Table 10 Characteristics of Commercially Manufactured Polycrystalline PSZ

Source	Stabilizer	Density (g · cm^{-3})	Flexural Strength[a] (MPa)	Fracture Toughness[a] (MPa · m$^{1/2}$)	Young's Modulus of Elasticity (GPa)
Australia	MgO	5.75	700	8–15	200
Japan	MgO	5.75	900	4–7	201
Japan	Y_2O_3	6.03	980	9	210
West Germany	Y_2O_3	5.77	650	7–15	210
U.S.	MgO	5.60	430	4.7	192
U.S.	Y_2O_3	5.73	575	2.3	200
U.S.	Y_2O_3	5.24	200	2.8	170
U.S.	MgO	5.08	300	6.0	162

[a]Tests performed at the Naval Research Laboratory, Washington, D.C.

Source: Ref. 21.

Germany have produced PSZ with substantially higher values for both strength and toughness than the currently available U.S. PSZ.

Research in structural ceramics in West Germany has been funded at a level of nearly 100 million marks over the past 10 yr. This research, which began in 1974, was sponsored by the Ministry for Research and Technology (BMFT) for the program dealing with ceramic components for gas turbines, with almost half the funding provided by industry. This current program ends in 1983 and a program called "Car for the Year 2000" has been canceled. Currently, the German ceramic industry and some sectors of the government are attempting to form a program much like the proposed DOE ceramic program.[22]

The Japanese government began to fund structural ceramic research comparatively late, in 1978. Yet the Japanese are now regarded as ahead of, or at least even with, the U.S. The first ceramic program was a part of the "Moonlight" Project, which dealt with five energy technologies (one of which was advanced gas turbines). More recently, the Ministry of International Trade and Industry (MITI) has begun funding research for advanced materials, which include what the Japanese term "fine" or technical ceramics. This project is a part of a larger program called the Next Generation Industrial Base Program. A number of smaller government programs are also funding research in structural ceramics. The myriad of programs, as well as a number of interpretations of their efforts, has made estimation of their level of funding difficult. However, an estimate in excess of $60 million over the next 10 yr appears conservative. Furthermore private sector ceramic research has been estimated at 10 times the government level.[23]

In addition to the substantial government funding for research in ceramics, the Japanese are generally acknowledged to have three other advantages in vehicle manufacturing: (1) a historical lead in small car/engine technology, (2) relatively low labor costs, and (3) good labor-management relations. Breakthroughs in ceramic technology, combined with these advantages, would further enhance the Japanese position relative to that of the U.S. in the auto industry.

Two recent articles in a Japanese daily newspaper, *Neihon Keizai Shinbun*, support the U.S. concern about the effects of Japanese research in ceramics. First, Toyota announced that it would be using ceramic-aluminum inserts in the pistons of their 2.2-L diesel. This is the first time that ceramics have been used in the mass production of heat engines; more widespread applications may follow. The second article describes and assesses a prototype ceramic diesel engine that has been built by Kyoto-Ceramic. Translations of both of these articles are included in Appendix B. In addition, Appendix C contains a synopsis of *American Counterattack - Plan Z, 1985*, a novel published in July 1982 that had some influence on Japanese public opinion about ceramics research. (Appendix B, second newspaper article, contains a description of the impact of this novel.)

3. The Need for Federal Research Support

3.1 ASSESSMENT OF PRIVATE SECTOR SUPPORT

Given the large, though potentially long-range, market for structural ceramics in the heat engine applications, it is essential to examine the need for any government support for research. Given a well-defined free market, it would seem reasonable that industry would do whatever is necessary (including research) to maximize profits on a very promising technology. However, contacts with the industry have indicated that without significant government support, much, if not all, of the long-range research would be abandoned. Why then is the private sector hesitant to perform this work on its own?

The issue of sufficiency in private-sector R&D is complex. First, the potential markets are well defined for structural ceramic applications, but the best technical approach (i.e., which ceramics are best) and the time frames are not, especially for the heat engine uses. The impetus for more fuel-efficient engines has been the petroleum supply interruptions and consequent fuel price increases during the 1970s. If world oil supplies continue to be used at a consistent rate, fuel prices will probably show only a moderate rise (i.e., at or below the rate of inflation). With reasonably adequate fuel supplies for industry, the incentives for research into alternative fuels or more-efficient engines will be diminished. The long-range demand for advanced engines -- the adiabatic diesel, gas turbine, or Stirling engine -- will, consequently, be very low, and there will be little, if any, return on a research investment.

Compounding that risky aspect of the research is the fact that the gasoline and diesel engines have also been improved significantly, thus minimizing the need for advanced engines. Before the fuel crises of the 1970s, high-mileage cars provided about 30 mpg. The 1983 Environmental Protection Agency (EPA) test cycle results ranked a Volkswagen Rabbit diesel as the top fuel-economy car with 50 mpg city rating (67 mpg highway rating). The highest-mileage gasoline car was the Honda Civic with a 46 mpg city rating. Chevette and Pontiac 1000 were the highest-rated U.S. cars, each with 42 mpg.[24] These advances in existing engine and car designs make the eventual penetration of ceramics more difficult. In addition, the eventual return on the research investment is further diminished with the poor sales cycles that fuel-efficient cars periodically encounter.

The ceramic industry has additional reasons to hesitate to invest in structural research. Even if the timing of structural ceramic commercialization could be accurately predicted and the market for fuel-efficient vehicles grows, the industry may perceive that foreign competition (with government support and funding) has committed greater resources than the U.S. industry alone would be willing to do. If the U.S. industry research effort has little or no government support, then there is a very low probability that the U.S. could have commercial products available as quickly as a foreign competitor that has both industry and government funding. Under these circumstances, a company deciding to commit funds for research could easily find itself beaten by a foreign firm achieving earlier results and dominating the market -- their governments having "bridged the gap" in research funding. The limited number of research dollars available suggests investment where there is the highest probability of a return-on-investment, i.e., a short-term marketing approach. Furthermore, if industry does attempt the research alone, the extensive U.S. antitrust legislation would preclude intraindustry cooperation in the research effort, thus further reducing the probability that an industry research effort, exclusive of government support, could cost-effectively compete with either Japan or Germany.

Such a scenario implies that the U.S. ceramic industry would be willing to forego all of the potential benefits of the technology in the attempt to reduce the long-term risks. As will be discussed in the next section, the ceramic industry could not expect to capture all of the benefits that would accrue from structural ceramic commercialization. Many of the benefits (in terms of profits and market share) will go to the users of the ceramic components, e.g., vehicle manufacturers, who may or may not have invested in the research. Industries, and subsequently the employees and stockholders, that utilize advanced ceramics in their processing will benefit from the increased productivity. Consumers that are offered more-fuel-efficient vehicles will secure some of the benefits. The defense industry in using structural ceramics will obtain the benefits of higher performance and more-reliable weapons systems. Consequently, the social rate of return is likely to be much higher than the private rate of return for the research investment in ceramics.

With the significant advances made by the U.S. in structural ceramics during the last decade and the intense competition that is developing in Japan and Germany, this would seem to be the wrong time to decrease the level of U.S. research. Four additional reasons for private industry hesitation to solely fund advanced structural ceramic research are a function of the current times and affect all U.S. industry research.

1. The current high interest rates make borrowing expensive and reduce the funds available for research.

2. The current decline in profits (and, in some industries, mounting losses) make it difficult to fund the research without borrowing.

3. The current low stock prices make it difficult to raise capital through the stock market.

4. The length of time that a new product remains state of the art is becoming shorter, so there is less opportunity to capitalize on the research expenditures.

3.2 RESEARCH AND DEVELOPMENT: INVESTMENT VS. "BORROWING"

Two approaches can be used for the commercialization of new discoveries. One approach stresses basic research followed by market diffusion and utilization. The other approach emphasizes the use of concepts developed by other research groups (i.e., simply "borrowing" the discoveries) for the diffusion and utilization stages. Recent history, from World War II until the 1970s, shows that the Japanese have very successfully adopted this latter strategy. This section explores these two approaches and looks at the development of high-temperature, structural ceramics in this context.

3.2.1 Productivity and Basic Research

Historically, most economists were pessimistic about the limits of per capita output. In their view, population would expand whenever the standard of living rose above the subsistence level, with the result that more and more people would be working a relatively fixed amount of land. Given the law of diminishing marginal returns, it appeared certain that per capita output would eventually fall. That this decrease in output did not happen is largely the result of technological change. Innovations have more than offset the law of diminishing returns and allowed the levels of food supplies, transportation, etc., to increase faster than the population level.

Economic growth is due largely to increased productivity (output per unit of input) and the rate of productivity growth depends largely on the rate of technological change. However informal or structured it may be, R&D initiates the forces for that technological change. Recently, economists (Mansfield,[25] Griliches,[26] and Terleckyj[27] among others) have been exploring the relationship between basic research and growth of productivity. Mansfield[28] empirically shows that a firm's number of innovations during 1951-1971 was highly correlated with the R&D expenditures for basic research. Further, Mansfield's[25] and other studies indicate that the average social rate of return (measured by society's increased standard of living) from investments in new innovations tends to be high. Empirical studies of these concepts have begun appearing in economic journals relatively recently. As a result, the relationship between R&D expenditures and innovation has been explored, but the exact causal relationships have not yet been established. Microeconomic studies of the return from particular innovations generally estimate the social benefits from innovation using the following model. If innovation results in a shift downward in the supply curve (from S_1 to S_2 in Fig. 7) for a product, the area under the demand curve (D in Fig. 7) between the pre- and post-innovation (from P_1 to P_0) supply curves measures the social benefit. If all other prices remain constant, this area equals the social value of the resources saved as a consequence of the innovation. This is shown by the area P_0P_1ABC in Fig. 7. Thus, economists have found that investment in basic research is positively correlated with the number of innovations, which in turn results in increased productivity for society, leading to a positive

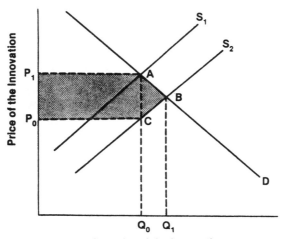

Fig. 7 Increased Social Value of Innovation (P_1 and P_0 = pre- and post-innovation; Q_0 and Q_1 = pre- and post-quality; S_1 and S_2 = pre- and post-supply)

social rate of return. In fact, Mansfield[25] indicates that the marginal social rate of return is generally at least 30% for a new technology. Thus, investment in basic research can have significant benefits to the U.S.

3.2.2 The Relationship between Basic Research, Innovation, and Diffusion

Most economists visualize three main stages by which technological changes occur: invention (basic research), innovation, and imitation or diffusion. At the invention stage, the new insights underlying a product or process are achieved and subjected to preliminary testing. At the innovation stage, the match between technical possibilities and market demand or user needs is recognized and the invention is developed to full-scale production and introduction into the market. In the diffusion or imitation stage, other organizations observe the success of the innovating organization and imitate the products or processes. An important extension of this concept is the argument that society's ultimate interest in science and technology is through its role in both the introduction and spread of new and more-effective technology and improved standards of living. If diffusion and utilization are swift, then the social value of science and technology investments is likely to be high. Diffusion has three closely related consequences: wider utilization of the innovation, the erosion and eventual elimination of the profits initially earned by the innovating organization, and a transfer of the main benefits of the innovation from the innovator and early imitators to consumers or users through price reductions or better products. The scope of the imitation hinges on the presence and strength of barriers (supply networks, patents, communication, etc.) to diffusion. Government-sponsored research in private industry poses a unique paradox. On the one hand, wide and rapid diffusion of the basic research breakthroughs and resulting innovations increases the social rate of return but also speeds the erosion and eventual elimination of proprietary profits, which lessens the incentive of private industry to undertake the basic research. Thus, it is important for industry and government to accurately project their benefits before negotiations for joint sponsorship of research is undertaken. Extreme care must be taken to protect society's interest and, at the same time, provide proper incentives for private business.

3.2.3 A Question of "Borrowing"

Little definitive research has been done on the benefits of either undertaking basic research or "borrowing" the resulting technology at the diffusion stage. Certainly there are positive social benefits in "borrowing" technologies, as increased utilization results in lower prices and improved quality to the consumer. Mansfield's work[29] indicates that the social return from basic research is about 30%. This percentage reflects, in part, the early monopoly rent (profits when no competitors exist) of being the "first" society to have the discovery plus the rent in the diffusion stage. Thus, the expected social return on basic research is higher than the return on borrowing (diffusion rents only). The key question, however, focuses on the difference between these returns and whether the disparity is sufficiently large to warrant government sponsorship of a program.

3.2.4 Ceramics: A Preliminary Look

Figure 7 shows that the additional social value created by basic research comes from the shift in the supply curve downward and to the right. The U.S. DOE, as well as other federal agencies, has provided incentives for various research projects with this concept in mind. At present, it has been proposed that DOE sponsor a base technology ceramic research program. Assuming that the social return on ceramic research is close to the average, the citizens of the U.S. could expect an enhanced standard of living through new or improved vehicle engines as well as commercial and industrial products that will contribute to U.S. productivity. There are several ways in which the ceramic research is likely to have positive social value. The following points illustrate how the supply curve in Fig. 7 can be expected to shift from S_1 to S_2.

- Major developments in ceramics will create new markets and expand old ones. These markets include vehicular transportation (turbochargers, truck engines, tracked vehicle engines, auto engines), aerospace (aircraft engines, missiles, sensors linked to electronic control systems) and turbine refractories (heat exchangers). The extent of the penetration of ceramics into these markets will depend on the reliability and cost-effectiveness of the ceramic components in these technologies. If the components do not appear reliable, the market penetration will not occur even if the price is low. But if reliability is demonstrated (as it has been in numerous product applications), significant penetration will occur and the value to society from the new and improved products may approach or even exceed the 30% estimated by Mansfield.

- Advances in ceramics will improve the efficiencies of several alternative forms of transportation, resulting in less energy use and less dependence on foreign sources of transportation fuel. The improved efficiency of heat engines will initially lead to higher-temperature operation. Without ceramics, some of these efficiency improvements will probably never be realized. Fuel efficiency improvements can be expected as ceramic components are used in contemporary diesel engines, then later in the adiabatic diesels and the gas turbines.

- The dependence of the U.S. on imported critical materials (particularly cobalt, nickel, chromium, and tungsten) will be lessened because high-temperature ceramics can substitute for superalloys that contain many critical materials. In general, many scarce and/or strategic materials are candidates for replacement by ceramics in wear, corrosion, and high-temperature applications. Today, silicon carbide is displacing tungsten carbide with high cobalt content for seal rings. It is projected that alloys containing tungsten and titanium will be replaced in cutting tools and abrasives; beryllium will be replaced in reinforced spars and tiles; and nickel, chromium, platinum, and palladium will be replaced in anticorrosion piping, valves, and seals; many other replacements are possible. The decreased U.S. vulnerability resulting from the lower requirements for these strategic materials would lessen the need for the government to stockpile these resources. The funding released by lower stockpile limits and by sales of surplus materials could be considered an offsetting benefit to the cost of a ceramic research program and would be another indirect benefit to U.S. citizens.

One final point in the evaluation of the developing vs. borrowing concept for advanced ceramic technology concerns Japan. It is well recognized that Japan is the principal U.S. competitor in this field; their exact position relative to U.S. in the race for ceramic development is a subject for debate, but their recent long-term commitment to ceramic research is a matter of record. Japan, however, which has long been content to borrow U.S. and European technology (e.g., the transistor, robotics, etc.) and capitalize on the commercialization, is no longer willing to do so. Increasingly it has become important for them to do their own research and reap the early monopoly benefits discussed earlier.[30] This development in itself, would provide at least *prima facie* support for the need for a U.S. long-term ceramic research program.

4. Industry Perspectives

Individual meetings were held with representatives from the ceramic industry (including ceramic manufacturers, engine manufacturers, and researchers), to informally discuss the economics of high-temperature, structural ceramics. Some of the companies gave additional thought to the issues and responded after the meetings.

The economic issues can be broadly categorized as (1) the market potential for advanced ceramics, (2) the need for research and development, and (3) the threat of foreign competition. Perspectives varied widely among companies (even among individuals within some companies) on some of the issues. On other issues there was considerable agreement. The following question-and-answer format is used to convey the essence of the industry perceptions about the major economic issues.

4.1 THE MARKET POTENTIAL FOR STRUCTURAL CERAMICS

- What are the potential applications for structural ceramics?

Many potential applications exist for new, high-performance ceramics. In some markets, there is already limited production of these materials for use as abrasives, wear parts, and electronics components. For longer-term applications, the eventual commercialization depends on the commitment of an extensive research and development program, whether in the U.S. or abroad. The following list of potential markets is divided into intermediate-term and long-term time frames, although some disagreement exists about the timing of commercialization.

<u>Intermediate-Term (by the year 2000 including improvements in some existing uses)</u>

- Cutting tools

- Wear parts (e.g., bearings, seals, valves, nozzles)

- Integrated circuit substrates

- Anticorrosion parts (e.g., valves, seals, piping, coatings)

- Air-fuel sensors

- Glow plugs

- Catalyst honeycombs

- Diesel wear parts (pushrods, tappets)
- Diesel combustion parts (cylinders, piston caps, valves)
- Turbocharger rotors
- Turbine static parts (combustors, shrouds)
- Missile engine components
- Radomes and infrared domes for missiles
- Heat exchanger components
- Turbines for small power-generation units
- Stationary heat engine parts (for diesel and Stirling engines)
- New refractories (kiln furniture, heating elements)

Long-Term (from 2000 to 2025)

- Improved operations for the intermediate-term applications
- Turbine rotating parts (rotors, regenerators)
- Battery and fuel cell components
- Solar concentrator targets
- Magnetohydrodynamic generator components
- Fusion reactor components
- Turbines for large power-generation systems
- Engines for manned aircraft
- Advanced tank engines

- How will the costs of the structural ceramics compare to the costs of the conventional materials that they will displace?

Initially the new ceramics are likely to be more expensive than the current materials, except in cases where the current materials are already expensive alloys, such as tungsten carbide or tungsten cobalt in the wear parts application. The current high cost of advanced ceramics is due in large measure to the low production levels and the consequent lack of any economies of scale. The high cost of starting into production is probably similar to the capital costs associated with aerospace superalloys. In time, with advances in processing and mass-production techniques, the fabrication costs will likely be comparable to those for competing materials, especially the superalloys. The competitive position for structural ceramics will be further enhanced by any shortages of strategic materials (see the discussion below) for which the U.S. is dependent on foreign sources.

Whereas the capital (or production) costs are expected to be comparable to those for existing materials, the operating costs, although varying by application, are likely to be 30% less for the ceramic components. Also, the advanced ceramic materials are likely to achieve better performance than conventional materials, through longer life, greater reliability, or lower total life-cycle costs. In addition, there are some applications (such as gas turbines for cars) that would never be commercially viable without the use of structural ceramics because no effective alternative material is available.

- To what extent will the proposed ceramic research program affect the market penetration of structural ceramics?

The proposed research program was perceived by the ceramic industry as significantly accelerating the commercial introduction of new components and consequently increasing the market penetration of ceramics into each of the potential applications. The current technological barriers (see discussion below) are well recognized. Based on the past experience of the rate of technological advancement in ceramics (which depended heavily on federal funding), it is generally agreed that a substantial, concerted program will be required to bring ceramic technology to the point of commercialization. It was also generally agreed that any long-term application would not be achieved by the U.S. industry without a significant funding commitment by the federal government. With reduced federal funding, more opportunity would be present for foreign ceramic fabricators.

If a coordinated research program is established relatively soon, the market potential for ceramics in many applications will be greatly enhanced. For near-term applications, the research program will not affect the early market penetration, but by the year 2000 the benefits of new materials and processes will strengthen those markets. On the other hand, the program will likely produce a doubling or tripling in the market penetration by the year 2000 for the long-term, high-risk applications. Table 11 summarizes the range of market penetrations for selected applications as assessed from the meetings with the representatives from industry.

Table 11 Structural Ceramic Market Penetration

Application	Market Penetration (%) by 2000	Market Penetration (%) by 2025	Impact of Proposed Program
Transportation			
Diesel engines[a]	5-20	up to 50	Substantial
Turbocharger parts	10-75	up to 80	Appreciable
Missile engine parts	5-50	up to 50	Appreciable
Cutting tools	5-10	10-25	Appreciable
Wear parts	5-30	15-40	Moderate
Stationary turbine/diesel			
Small power generator	5-30	10-60	Appreciable
Large power generator	0-5	10-20	Appreciable
Heat recovery systems	50-70	75-95	Appreciable

[a]Combustion parts only; wear parts would be a higher penetration but less affected by the proposed program.

- What strategic materials can structural ceramics be expected to displace?

Advanced ceramics are candidates to replace a wide variety of strategic materials, especially those used in superalloys. Strategic or critical materials are (1) those for which the U.S. is heavily dependent upon foreign sources, (2) those for which the major natural deposits are located in communist countries or countries with potentially unstable governments, or (3) those that are simply scarce. As discussed earlier, these strategic materials include tungsten, cobalt, nickel, chromium, manganese, rhenium, boron, beryllium, titanium, platinum, molybdenum, hafnium, columbium, vanadium, and tantalum. In some applications, this displacement has already begun, e.g, the use of silicon carbide instead of tungsten carbide for seal rings. As the ceramics are substituted for these strategic materials, these critical materials will then become more readily available for other essential applications. In some applications, structural ceramics will not actually replace strategic materials, but will enable new products to be fabricated that were not previously feasible due to the limitations of existing materials.

4.2 THE NEED FOR RESEARCH AND DEVELOPMENT

- What are the crucial, technological barriers impeding the commercialization of structural ceramics?

Much of the problem that restricts the economic attractiveness of structural ceramics stems from the brittle nature of the materials, which in turn affects design and mass-production considerations. As a result, considerable research is needed to improve the mechanical and physical properties and fabricability of the materials. The following are areas for reseaarch: (1) controlling defect size, concentration, and distribution; (2) improving raw material processing, forming technology, green-state characterization, and firing and finishing techniques; (3) developing cost-effective, mass production processes; (4) identifying nondestructive evaluation techniques for quality control monitoring during production; and (5) increasing the understanding of ceramic characteristics by designers and users. Research in these areas will produce materials that are more reliable, tougher, and more resistant to thermal shock, contact stress, and environmental effects than are present materials. To be effective, the research program should be iterative between material development/component fabrication and the rig or engine testing.

- What are the competing technologies?

A number of technologies were identified that would or could eventually compete with structural ceramics. These include both cooled and uncooled superalloys, welded or machined superalloys, coated metals, coated carbon-carbon, oxide-dispersion-strengthened alloys, and advanced plastics.

In the heat engine application, the competing materials will be focused on the cooling system because extensive use of ceramics can eliminate the need for cooling. For most engineering applications that require the characteristics of ceramic material (high-temperature strength, low thermal expansion, corrosion and wear resistance, and high elastic modulus), there is no material with properties that can match those of the ceramics. The competing materials, especially the superalloys, are not likely to have performance and durability comparable to those of ceramics at high temperatures. In addition, the superalloys, which contain critical materials, are expensive. The turbine disks that support the blades contain the largest percentage of superalloys. In new designs, these parts can be totally replaced with ceramic parts.

- Are the current federal research funds for ceramics adequate considering the market potential?

Nearly all the firms expressed a general feeling that funding levels were inadequate. Only one indicated sufficient funding and qualified that response by indicating the adequacy was only until the markets were better defined. Most companies felt that current U.S. government support is less than that given by the governments of the major foreign competitors. In addition, the U.S. funding appears to be declining while foreign support is increasing. Furthermore, as there is currently no national coordinated plan that links basic and applied research, even if the foreign competition spent less, their programs would, presumably, be more cost-effective and yield results sooner.

Past federal funding was criticized for its short-term nature, with year-to-year funding uncertainties. Flexibility within the programs has been minimal, i.e., the need to meet component deadlines has allowed relatively little effort to be devoted to material development or to problems that have arisen during the programs.

The long-term benefits of structural ceramics (such as improved efficiencies in transportation and industrial applications, reduced dependence on strategic materials, and increased productivity) extend far beyond the companies involved in ceramics. Therefore, higher levels of federal funding are considered to be justified on the basis of these benefits. However, a well-planned program is required, with close interaction among government, industry, and universities. At least one company considered a federal program of $15 million per year for five years to be inadequate when compared either with the major foreign ceramics research programs or the expected benefits from such a research effort.

- How would the proposed federal ceramic research program affect industry's research efforts?

From discussions with people in the ceramic industry, it appears that all of the firms receiving federal support also spend company funds for internal research and development. The average percentage of company funds expended as a percentage of federal funds received appears to be about 50%, with a range of 15 to 75%. Also, the federal support appears to serve as a catalyst for industry research funding; therefore, a withdrawal of the federal support would probably prompt some firms to eliminate their research in this area altogether and others to reduce their spending to modest levels in order to produce near-term benefits. A few, perhaps, could be expected to continue their long-term research.

The most significant disagreement among companies in the industry concerned direction and control of the program. The split was generally between ceramic manufacturers and engine manufactures. The ceramic fabricators usually consider themselves as suppliers or "job shops," in which they make component parts to the specifications of their customers. They view a base technology research program as enabling them to expand beyond this short-term orientation and providing them with both direction and support for long-range work. On the other hand, engine manufacturers expect the program to be component-oriented, with the user of the component needing to maintain control over the direction of the research. They feel that otherwise the research would be an end in itself. Obviously, close coordination between supplier and user is required and resolution of this difference of opinion is a necessity for the successful implementation of the program.

Most companies seemed very interested in cost-sharing in a long-range research program if they could retain proprietary and data rights.

Cost-sharing is an area for negotiation and in at least one case would require a creative approach for a firm whose company policies prohibit cost-sharing.

- Is federal funding desirable for ceramics research and if so, why?

The long-term nature of the research with its concomitant high risk and the intense, government-supported, foreign competition were the most frequently cited reasons for wanting government support. These reasons with the others listed below provide a compelling argument for federal funding. The structural ceramic industry, which was estimated by one company to have sales of $300 million (excluding ceramic packaging), is too small to support the kind of long-range, research program that is needed. An average corporate R&D budget of 5% would yield $15 million/yr, but three-fourths of that amount would be needed for development and applications, not basic research. The remaining amount would likely be spent for research with only near-term applications. Thus, the industry is undercapitalized to perform long-range research for which there is no assurance of success or where there is a market demand that may change due to unforeseen circumstances, such as stable fuel prices.

As ceramics could replace strategic materials, improve productivity in many industries, and reduce fuel consumption, the benefits from their development are widespread. Federal involvement would, therefore, be logical because these benefits involve concerns of national interest.

4.3 THE THREAT OF FOREIGN COMPETITION

- Which countries are the principal foreign competitors?

Japan and Germany, in that order, are the principal foreign competitors. One company indicated that without a significant U.S. research effort these two countries could account for 70% of the U.S. market for structural ceramics. Other countries are not expected to pose a significant competitive threat. The Japanese already have 70-80% of the ceramic electronic packaging market. Because this electronics market developed rapidly, the commercialization of heat engine structural ceramics could possibly occur more rapidly than is generally supposed. With Japanese research in this area increasing and the U.S. effort declining, the Japanese have positioned themselves for such a possibility, while the U.S. has yet to do so.

- What is the position of the U.S. relative to the other countries in structural ceramic research?

A variety of opinions exist in this area, especially with respect to the relative position of the Japanese. Some in the U.S. ceramic industry view the current Japanese research and understanding of structural ceramics as even with the U.S. at the present time, while others see the Japanese as having moved ahead. There is consensus, however, that the Japanese have the momentum, as shown in Fig. 8.

The Japanese are ahead in electronics packaging and probably in zirconia-based ceramics. The Japanese are also ahead in establishing a long-range research program -- marshalling the human and material resources needed to develop high-performance ceramics. They also have a lead in exploitation and commercialization of products. The U.S., however, is probably the current leader in basic research and materials development. The U.S. also has more experience than any other country in the design of engine components and evaluation on test rigs. However, without a substantial U.S. ceramic research program for the next decade, the U.S. will be far behind Japan and Germany, with little hope for regaining the technological lead.

- What are the likely consequences of the U.S. failing to maintain a lead in structural ceramic research?

Some companies feel that it is highly probable that the U.S. ceramics industry would be essentially excluded from any of the "high tech" ceramic markets and, with few exceptions, be left with the markets for traditional ceramic products such as bricks, glasses, and porcelains. Companies that would incorporate ceramic components into their products (e.g., auto and aircraft manufacturers) would either have to import those components or produce an inferior, but probably more expensive, product. In either case, they would have lost important lead time and, in turn, would face a declining market share. The increase in imports would obviously have an adverse effect on the U.S. balance of trade.

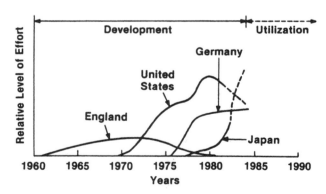

Fig. 8 A Generalized Comparison of Structural Ceramics Activity

Conflicting opinions exist as to whether the economic decline associated with foreign dominance would be long lasting. If engine manufacturers are able to buy ceramic component parts from foreign sources soon after successful demonstration and production start-ups, then little direct economic effect will occur except within the ceramic industry. If not, the effects will be extensive in both duration and magnitude and affect both the civilian and defense industries.

The defense applications for structural ceramics (such as missile sensors, engine parts, and radomes) will be delayed until research has advanced. It would place the U.S. in an extremely vulnerable position to become dependent upon foreign supply sources for key materials in defense and weapons systems.

5. Conclusions

The preliminary analysis presented in this report concludes that the benefits associated with using structural ceramics in heat engines warrant substantial government support for a base technology research program to ensure their development. Commercialization of structural ceramics in heat engines will have significant direct and indirect benefits for both producers and consumers. Several nations (notably Japan, W. Germany, and the U.S.) have had well-funded, joint government and industry research programs. Continued government funding for a base technology research program, however, appears uncertain in the U.S. and Germany. The ultimate effects on the U.S. economy of whether U.S. firms or foreign manufacturers lead in the commercial production of structural ceramics are summarized below.

With assumptions of market penetration rates obtained through ceramic fabricators, engine manufacturers, and institutions involved in ceramic research, an analysis of macroeconomic effects showed the differences that could be expected from two different scenarios, one in which the U.S. leads in structural ceramic commercialization and the other in which Japan leads. Peak year impacts (which are generally in 1995 and 2000) occur when one nation maximizes the sale of cars and trucks with ceramic engine components and just before international competition for those markets begins to erode the market share of the innovating nation.

The major economic effects include the following:

- Gross National Product. In a U.S. Dominant scenario, the GNP was expected to be $28 billion (1981 $) higher in the peak year than in a Base Case forecast, or 0.6% higher than what could have been expected without the commercialization of ceramics. Conversely, in the Foreign Dominance scenario, GNP was projected to decline by $11 billion or 0.3% compared to the Base Case. In cumulative terms, real GNP was higher by $280 billion over the period 1985-2005 in the U.S. Dominant Case. In the Foreign Dominance Case, cumulative real GNP was projected to decline by $110 billion for the same period.

- Employment. Peak year employment was forecast to increase by 250,000 jobs in the U.S. Dominant Case, reducing the projected unemployment rate from 6.5% to 6.3%. In the Foreign Dominance Case, the importing of half a million foreign cars with ceramic engine systems resulted in a loss of over 100,000 jobs in the worst year.

- Balance of Trade. The use of structural ceramics in heat engines affects the balance of trade in two major areas --

imports of petroleum and imports of autos. In the U.S. Dominant Case, fuel savings continue as sales of new cars and trucks equipped with ceramic engines yearly increase the total size of the fleet that has the more efficient engines. Domestic production of these vehicles reduces the number of imported cars, which reaches a low point in the year 2000. Total import savings approach $28 billion (1981 $) in 2005. In the Foreign Dominance Case, the value of the imported cars exceed the value of the fuel savings so that there is a negative effect on the balance of trade, reaching $5.5 billion in 2005.

- Fuel Savings. Fuel savings are achieved under both scenarios, although the savings are greater for the U.S. Dominant Case. The greater energy savings are due to two assumptions: (1) broader market penetration (into stationary and heavy-duty engines for the U.S. case) and (2) a legislative limitation on the percent of imported cars (30% of new car sales) under the Foreign Dominance Case. Annual fuel savings reach 1.1 quad by 2005 in the U.S. Dominant Case and 0.3 quad by 1995 (the peak year) in the Foreign Dominance Case.

These ranges of effects on the national economy are principally the result of structural ceramic applications in vehicular engine systems and, to a small degree, of ceramic use in stationary power generation systems. It is recognized that a base technology research program in structural ceramics will, in addition, have a substantial influence on other market areas for advanced ceramics including aerospace, weapons systems, electronic, optical, and refractory applications. Because these additional product markets are likely to be quite important, spin-off applications from the structural ceramic research for heat engines may be as important to the national economy as the direct benefits.

Reliable ceramics in engines also have the potential for reducing the nation's dependence on a number of strategic materials. These materials include beryllium, cobalt, chromium, columbium, manganese, nickel, platinum, tantalum, titanium, and tungsten. All of these materials are stockpiled by the U.S.; for most of them, domestic reserves are so small that imports account for 90-100% of the supply. The major sources for some of the materials (e.g., the Union of South Africa and the U.S.S.R. for chromium and platinum and Zaire and Zambia for cobalt) represent potentially unstable suppliers over the long term. Alloys of these materials are used in a variety of applications including heat engines, wear parts, chemical ware, abrasives, tiles, and castings. Advanced ceramics could replace these alloys in many applications. One indication of the magnitude of recent expenditures is that a $100 million appropriation for the U.S. stockpile acquisition program was spent entirely on cobalt. Though this analysis provided only a cursory examination of the strategic materials issue, it is conceivable that the cost of a base technology research program in ceramics could be justified on the savings of imported strategic materials alone.

Although substantial benefits will obviously be derived by the commercialization of structural ceramics in heat engines, this analysis points out that private industry may be hesitant to totally fund the basic research necessary to ensure commercialization. Although the potential markets are well defined, the eventual ceramic materials and processing techniques are not yet known. In addition, the time-frame for commercialization is still not well described. The principal markets for the structural ceramics are the gas turbine engine and the adiabatic diesel. However, the impetus for advanced, more-fuel-efficient engines has been rising fuel prices and unstable supplies; now fuel prices are declining with the result that there is less pressure for R&D for advanced engines. All of these factors indicate that structural ceramic research is a long-range, high-risk endeavor, an area where industry would be reluctant to concentrate its limited research funds. The widespread perception that other governments (particularly Japan) have committed substantial funding for multiyear programs in ceramic research is further cause for industry's hesitancy to proceed without government support.

This preliminary assessment, while relatively optimistic about overcoming the technological barriers, is very conservative in the assumptions regarding market penetration of structural ceramics. Nevertheless, this generally conservative analysis is strongly supportive of establishing a base technology research program in structural ceramics, jointly funded by government and industry. It is recognized that the lack of an engineering-cost analysis, due to the short time-frame for conducting this study, is a limitation in giving unqualified support. Until such a detailed engineering-economic evaluation is undertaken, the conclusions reached in this preliminary study cannot be confirmed.

Appendix A
Changes in the Economic Model

A.1 THE DRI ANNUAL MODEL OF THE U.S. ECONOMY

The DRI annual model is a macroeconomic model that has evolved from the nationally known DRI quarterly macromodel of the U.S. economy. It is designed for long-run analysis of the U.S. economy in cases where excessive details are not required. The most significant long-term issues from the quarterly model have been expanded in the annual model (e.g., greater detail has been provided for energy and demographics in the annual model). In contrast, the variables that are crucial only for short-run effects have been suppressed. Moreover, certain concepts are treated at a higher aggregation level in the annual model than in the quarterly model.

Instead of the 1207 variables in quarterly model, the annual model has 266 economic variables (75 exogenous and 191 endogenous variables). All of the eliminated variables represent disaggregated detail. The annual model is an econometrics model with the basic structure of the model based on well-known and generally accepted economic theories. It has 191 equations, 142 of which have simultaneous relationships capturing the interaction in the economy. The model's treatment of economic principles has been well-scrutinized by the economics profession over the last 10 yr.

The major sectors of the model deal with demographics, labor force, energy, consumption, investment, housing, trade, government, prices and wages, and finance. The equations in the model have been econometrically estimated based on extensive historical data. The performance of the model has also been tested comprehensively (for equation error as well as dynamic simulation error) for the period 1966-1980.

A.2 CHANGES MADE IN DRI BASE CASE SCENARIO FOR THE U.S. DOMINANT CASE

1. Investment Spending - Industry was assumed to incur the extra investment required to produce ceramic heat engines, with the annual investment estimated for each of the market penetration levels of ceramics heat engines. Table A.1 shows the assumed annual investment; by 2005, the cumulative investment will approach approximately $6 billion (1981 $). (The change in investment spending was converted to 1972 dollars to match the model national income account data.) The model variable "investment of producer's durable equipment" was increased by this amount through an add factor.

2. Annual Fuel Savings - The fuel savings were derived on the basis that the efficiency of ceramic engines is expected

Table A.1 Assumed Investment and Resulting Fuel Savings of Ceramic Heat Engines for the U.S. Dominant Case Scenario

Year	Increase in Investment to Date (billions of 1981 $)	Fuel Savings/Yr (billions of gal of gasoline)
1985	0.13	0.0
1990	0.68	0.28
1995	2.01	1.37
2000	3.72	4.37
2005	5.95	8.32

to be 30% higher for mobile engines (transportation), and 35% higher for stationary engines (power generation, etc.) than the efficiency of conventional engines. The annual fuel savings, in billions of gallons of gasoline, for the U.S. Dominant Case is also shown in Table A.1. The direct ceramic engine fuel savings were used to adjust both the appropriate fuel demand variables and the consumer spending category downward.

3. Automobile Imports - In its long run outlook "Trendlong 2007B,"[31] DRI projects the share of import car sales to be 25.1% in 1982, 24.5% in 1990, 25.2% in 2000, and 25.9% in 2005. Under the U.S. Dominant Case, it was estimated that about 2.54 million of the cars sold will have ceramic engines in 2000. Further, DRI projects that new car imports will be 3.2 million under the Base Case scenario. It was assumed that the domestic producers using ceramic engines will capture a portion of the market from the imports (presuming a U.S. lead in ceramic technology). If the cars with ceramic engine components split the conventional market for domestic and imports cars equally, imports will be reduced by 1.27 million cars in 2000. The new level of imported cars will then be 1.93 million cars (about 15% of the new car sales in the U.S. during 2000). It is reasonable to assume, however, that Japanese companies will not sit idle and see their U.S. car markets dwindle indefinitely. Therefore, for the U.S. Dominant Case, it was assumed that the import share will not decline below 15% of all new car sales beyond 2000. For the years between 1990 and 2000, the import share was

assumed to gradually decline from 24.5% in 1990 to 15.0% by 2000, as indicated by the ceramic penetration curve. In the DRI model, the nonoil imports category was adjusted downward by using the lower levels of imported cars and the price of the car.

4. Retail Auto Prices – The total cost of cars with ceramic engine components was raised by approximately 2% to account for the technology development.

A.3 CHANGES MADE IN DRI BASE CASE SCENARIO FOR THE FOREIGN DOMINANCE CASE

1. Automobile Imports – Under this scenario, it was assumed that Japan would be able to sell 10,000 additional cars equipped with ceramic engines during 1985, the first year these cars were introduced. Sales of ceramic-engine cars were assumed to increase to a level of 100,000 cars annually by 1990. This increase in imports translates to a 25.2% share in 1990 for imported cars as compared to 24.5% share in the DRI Base Case scenario. The ceramic-engine cars were assumed to finally reach saturation level by 1995, when their sales volume reached 500,000 cars/yr. In 1995, this penetration level of imported ceramic-engine cars corresponds to an import car share of 27.8% vs. the 24.2% share under the DRI Base Case scenario. Beyond 1995, the foreign suppliers were assumed to maintain the import levels of 500,000 cars/yr. Cars with ceramic engines thus enabled the foreign suppliers to maintain their high level of current imports (26.2% in 1981 compared to 15.6% in 1973). The domestic automobile industry was assumed to be unable to recapture its lost share of the market. For the Foreign Dominance Case in the DRI model, the import spending was increased to account for the increased sales of imported cars.

2. Annual Fuel Savings – As for the U.S. Dominant scenario, the fuel savings were derived on the basis that the fuel efficiency of ceramic automobile engines will be 30% higher than the efficiency of conventional automobile engines. The annual fuel savings in billions of gallons for the Foreign Dominance Case are shown in Table A.2. These assumed fuel savings were applied to the transportation sector in the DRI model. The consumer spending category for energy services was also adjusted downward.

Table A.2 Imports of Ceramic-Engine Automobiles
and Resulting Fuel Savings for the Foreign
Dominance Case Scenario

Year	Imported Ceramic Engine Automobiles		Fuel Savings	
	Annual Sales (10^3 Sales)	Stock (10^3 Cars)	Billions of gallons of gasoline	Quad
1985	10	10	0.0	0.0
1990	100	330	0.05	0.006
1995	500	2,020	0.25	0.030
2000	500	4,200	0.46	0.055
2005	500	5,000	0.49	0.058

3. Retail Auto Prices - It was assumed that the cars equipped with ceramic engines will be 2% more costly. In the DRI model, the implicit price deflator for automobiles and parts was increased by 2% times the share of total sales. Because the share of ceramic-engine automobiles was less than 5% in any given year, the adjustment to price deflator for automobiles and parts was small.

A.4 CHANGES IN THE PROJECTIONS OF THE U.S. ECONOMY

As a result of the input changes to the DRI model, considerable change was projected to occur in the U.S. economy under both scenarios. A summary of the changes in several key indicators of the economy is given in Table A.3.

Table A.3 Ceramic Scenarios Compared to the DRI Base Case

	Change Compared to Base Case			
Scenario and Variable	1990	1995	2000	2005
U.S. Dominant Case				
Average Real GNP (billion 81 $)	2.5	15.2	28.2	16.9
Average Real GNP (%)	0.1	0.4	0.6	0.3
Average Inflation Index (%)	0.0	0.2	0.6	0.8
Industrial Output Index (%)	0.1	0.7	1.2	0.6
Unemployment Rate (%)	0.0	-0.1	-0.2	0.0
Employment Level (1000 jobs)	25.0	174.0	250.0	48.0
Total Imports (billion 81 $)	-0.7	-5.5	-14.3	-27.7
Fuel Import Bill (billion 81 $)	0.1	0.6	-1.2	-10.2
Foreign Dominance Case				
Average Real GNP (billion 81 $)	-2.2	-11.1	-6.5	-2.7
Average Real GNP (%)	-0.1	-0.3	-0.1	-0.1
Average Inflation Index (%)	0.0	0.0	0.0	0.0
Industrial Output Index (%)	-0.1	-0.5	-0.2	-0.1
Unemployment Rate (%)	0.0	0.1	0.0	0.0
Employment Level (1000 jobs)	-13.0	-106.0	5.0	4.0
Total Imports (billion 81 $)	0.8	3.3	4.5	5.5
Fuel Import Bill (billion 81 $)	-0.3	-1.8	-1.0	-0.6

Appendix B
Translations of Japanese Newspaper Articles

Translation by Koji Tsunokawa, Argonne National Laboratory

"TOYOTA UTILIZES CERAMIC METAL FIBER COMPOSITE FOR MOTOR PARTS"

Neihon Keizai Shimbun, July 28, 1982

Toyota announced on the 27th that it had succeeded in the technological development of mass-produced piston rings made of a ceramic metal fiber composite (a kind of Fiber Reinforced Metal - FRM) for diesel engines. Toyota will start using these rings in automobiles in late August. This composition or FRM has been used in the space shuttle and by the military industry. However, this is the first time, they claim, that the material is being used in the civilian field, for example in an engine, where cost is crucial. The material being used is aluminum reinforced with alumina silicate fibers. The material itself is made by a proprietary method in which a mold is coated with fiber, then the aluminum is poured into the mold to make the FRM piston part. They developed this technology with another Japanese company -- Art Kinzoku Kogyo -- whose stock is valued at 315 million yen (250 yen per dollar). The development took two years and nine patents are associated with it. Patent rights have been applied for in 10 foreign countries, including the U.S. The new piston inserts will last six times as long -- 300,000 km at full throttle -- as those made of nickel composite. The power of the engine can be raised by 5%, the noise level will be decreased, and an increase in fuel efficiency can be expected. Toyota has not yet made public the types of engines in which the new material will be used. Speculation, however, is that the new material will first be used with L-type engines, which are approximately 2000 cc in size and used in the Crown and Mark II models. (These are larger than the Amerian version, Corona.) It is probable that Toyota will also introduce this material gradually into truck and bus diesel engines. Ceramic fiber is widely used as an industrial material for preventing heat transfer, but it has been very difficult to combine this material with metals. This material developed by Toyota is one of the first cases in the world, since the only previous experience with fiber reinforced metal has been a kind of FRM with boron combined with aluminum in use for the space shuttle.

"THE BATTLEFRONT OF ADVANCED TECHNOLOGY DEVELOPMENT" (No. 5 in a series)

"GATHERING AROUND A BILLION YEN INDUSTRY"
Nihon Keizai Shimbun, August 20, 1982

Ceramic Engine Has Been Developed

At the Kyoto-Ceramic (Kyocera) Comprehensive Research Laboratory in Kokubu City, Kagoshima Prefecture, there is a commanding view of smoking Mt. Sakurajima to the south. A car stopped just in front of me -- the engine sounding somewhat noisy, though not unpleasantly so. Opening the hood, I was able to look at the peculiar engine. It did not have a radiator for water cooling or fins for air cooling. The engine looked totally different from the ordinary ones I was accustomed to. The simple round shapes of the cylinders were impressive.

This engine was the Kyocera ceramic engine, which succeeded in actually powering a car (a world's first) at the beginning of this year. The engine belongs to a fuel-efficient "dream car," and intense worldwide competition is now underway among manufacturers of automobiles and engines to achieve commercialization.

There is a continuous flow of visitors, both domestic and foreign, to the Kyocera Comprehensive Research Laboratory these days. Among these visitors are high-ranking officials from the U.S. Department of Defense, as well as engineers from the automobile and engine manufacturers.

All the parts of this engine that are suitable for ceramics have been replaced with ceramic materials (the cylinder, piston, combustor liner, and turbocharger rotor), according to Yoshimitsu Hamano, the head of Kyocera Comprehensive Research Laboratory.

Metals melt when heated to more than 800°C. Therefore, conventional engines made of metals need to be cooled by water or air. Ceramics are, however, heat-resistant as exemplified by their successful use in the outer wall tiles of the space shuttle. Ceramic material will withstand the combustion heat of automobile engines, which is more than 1000°C. Without cooling, the combustion energy can then be fully utilized. "The fuel efficiency of conventional engines can be improved no less than 30%," said Mr. Hamano.

Will There Be A Counterattack from the U.S.?

American Counterattack - Plan Z, 1985, (see Appendix C for synopsis) a near-future novel written by Katsuaki Sasa, is the talk of the town. The plot centers around the destruction of the Japanese auto industry, including Toyota and Nissan, by an American automobile manufacturer, presumably GM. It is the ceramic engine that will bring about the come-from-behind victory.

At Nissan Central Research Laboratory located at Oppama, Kanagawa Prefecture, I asked Mr. Michikazu Taguchi, the head of parts and materials division, whether an American counterattack can be expected.

The use of "ceramics as an automobile material is still limited because of problems with reliability and economy," said Mr. Taguchi, dressed uncharacteristically in working clothes. He continued, "Even with a successful test model, we will have to go a long way before we reach the technological level where only one out of 100,000 cars is defective. Cost consideration is also very critical. While one kilogram of iron costs only 10 yen, and 1 kilogram of plastic is 300 to 700 yen, one kilogram of ceramic costs about 10,000 yen."

When we parted, however, Mr. Taguchi added, "Ceramification is an irreversible tide. We are also building test ceramic engines and solving problems one by one." As a matter of fact, Nissan recently hired 10 veteran ceramic researchers and reinforced their research capability in ceramic engines.

The Chukyo area boasts of a long tradition in ceramic ware. This is another central place for the development of ceramic engines. After the successful building of a 50 cc all-ceramic, gasoline engine, Nihon Tokushu Togyo (the spark plug manufacturer) is now conducting a full load, proof-of-operation test of the engine. Its parent company, Nihon Gashi, which is located next to Nihon Tokushu Togyo, has succeeded in making an engine part on an experimental basis using a kind of ceramic called partially stabilized zirconia.

Toyota Central Research Laboratory is in Nagakuda Town, Aichi Prefecture. It is known worldwide for is success in the early 1970s in synthesizing Sialon, a kind of ceramic. "Although there is no prominent movement within Toyota, it is obvious that they are also conducting high-level ceramic research if you look at their patents in the area," commented Hiroshi Okuda, head of the Nagoya Institute of Industrial Science and Technology, Agency of Industrial Science and Technology.

Ceramics are regarded as the "third most important material" in human hands, following metals and resins. Although called ceramics, these materials are different from the material used in such simple things as china, which are made of sintered inorganic substances. The ceramics described here are of much finer structure, resulting from the sintering of substances in which impurities have been eliminated as much as possible. Therefore, they are also called fine ceramics or new ceramics. Raw material for ceramics such as quartzite and silica sand abound on earth. Compared with metals, their superior characteristics are that they do not melt or burn, are hard and not easily scratched, and are light. However, there are still many problems to solve such as how to overcome the fragility peculiar to ceramics.

"New ceramics will become a billion yen industry, with a half billion from automobile engines and another half billion from artificial bones and teeth," forecasts Ken-chiro Ando, managing director for materials of Toshiba Electric Co. This new market will be attractive to various manufacturers from the materials industry including chemical, fiber, cement, and steel companies, as well as automobile and machinery manufacturers.

Involvement of Military Strategies

In the light of the upsurge in the development of new ceramics on the part of private industry, MITI started the Research Project for Fine Ceramic Technological Development under a larger program called the Next Generation Industrial Base. "The objective is to develop the new ceramic industry in our country by establishing a technology comparable with that of the U.S. and European countries through the joint efforts of government and private sectors," said Mr. Nakajima Kunio, head of Fine Ceramic Division of MITI, his eyes sparking.

"The U.S. and European countries preceeded our country in launching large-scale projects in ceramic technology development with government subsidies." The U.S. DOE, in cooperation with NASA, has been developing new ceramic gas turbine engines for automobiles with a $100 million budget for the seven years from 1978 through 1984. The objective is to market the ceramic gas turbine engine for automobiles in early 1990s as a part of their strategy to decrease the U.S. reliance on petroleum.

In the U.S., especially under the Reagan administration, new ceramics attract attention because of their antiRussian, strategic importance. Chrome alloy is the only metal that withstands high temperatures of more than 1000°C. But most chrome deposits are distributed in Russia, eastern Europe, and other Communist countries. A Communist embargo could certainly cause "chromeshock" in free world nations. It may be that in the U.S. priority will be given to the development of jet engines for military and civilian airplanes and tank engines rather than for automobile engines.

In Europe, West Germany is also conducting a government-sponsored project involving auto manufacturers such as Mercedes, whose primary objective is the development of automobile ceramic gas-turbine engines.

"Japan is not inferior to the U.S or European countries in terms of the quality of material and the manufacturing technology of products. We also have many requests from foreign countries to produce goods on an experimental basis. Mass production technology is one of our specialties. Taking the initiative may not be impossible for us," said Yukio Fakatsu, chief engineer of Asahi Glass, after his recent trip to the U.S. and European countries.

The "New Ceramic War" between Japan, the United States, and European nations has been initiated by projects involving the governments and private sectors. With the possibility of an American counterattack, there is no question about the upsurge in the ceramic industry in the latter half of 1980s, which will be led by Japanese, American, and European auto manufacturers.

Appendix C

Synopsis of American Counterattack: Plan Z, 1985

by Katsuaki Sasa

A Synopsis of the Novel
Published July 10, 1982

Prepared by Koji Tsunokawa, Argonne National Laboratory

August 15, 1985, was a momentous day in Japan. For some time there had been rumors of a high-performance, fuel-efficient U.S. car that would be 20% cheaper than comparable Japanese models. The code name for this car was the Z car, and recently posters advertising the new car had begun appearing around Yokahama. Twenty ships, each bearing a Z car flag and loaded with new cars, had arrived at the U.S. military post in Yokahama. Numerous demonstrations, led by Japanese auto workers, had delayed the unloading for four days. However, pressure from the U.S. government (directly from President Wellington) to allow fair trade was effective, and on August 15, the process of unloading 50,000 cars began.

For a decade Japanese cars had flooded world markets. Their remarkable sales had been based on customer perceptions of high quality, low cost, and reliability. However, the world auto industry had been declining since 1982, whereas Japan was at the threshold of expanding its auto industry. Some observers saw Japan concentrating on improving production techniques while not being aware of the need to stimulate saturated car markets. In addition, the Japanese had concentrated on turbochargers as their major engine technology improvement. It was known that the Americans were spending their engine research efforts on ceramics, but that investment was considered by the Japanese to be too costly.

The unloading of the Z cars was quite a media event. Coverage was provided by both broadcast and print media. The television broadcast was especially dramatic — visible on the side of each car as it rolled off the ship was the name of the city for which it was destined. It was evident that the Z car would be available anywhere in Japan.

On the following day, August 16, there was a special taped television report describing the coming car war. One of the reporters, Mr. Nakagawa, interviewed Robert Jenkins, a vice-president for the M.G. Corporation, which built the Z car. While Nakagama was at home waiting to see the broadcast of the entire special report, he noticed a newspaper article that described the extent of the M.G. takeover of Japaneses auto dealerships. One hundred dealerships had been bought and closed down, only to reopen under the World logo (indicative of a World car), which was an M.G. subsidiary.

The special report was sponsored by Nichirin Mobile and Tokai Jidosha [presumably Nissan and Toyota]. The commercial messages were a virtual declaration of war against the U.S. cars. One company said that they had been selling world cars for years -- so what was new? The other company advertised their car as not just a world car, but a universal car.

A former Nichirin Mobile dealer, now selling World cars, described in the interview the fantastic capabilities of the Z car. He described the advanced instrumentation that allowed speed, rpm, fuel consumption, traffic information, route guide, and destination arrival time to be displayed at the driver's command in the center of the steering wheel. A radar device was combined with a voice synthesizer to produce unique information, e.g., "a truck is approaching from the rear at 60 kilometers per hour, distance 50 meters." An air conditioner was standard, which was unusual in Japan. But the most unique feature was the ceramic engine. Two sizes were available: 1500 cc and 2000 cc. Conventional engines could generate 80-100 hp, but the 1500 cc ceramic engine was rated at 135 hp -- comparable to a 2000 cc traditional engine. The 2000 cc ceramic engine was rated at 165 hp or equivalent to a 2800 cc turbo-equipped conventional engine. Water cooling systems were eliminated, resulting in 30% reduction in engine compartment size, which was then used to increase passenger compartment size. Fuel consumption was 18 km/L and 15.5 km/L for the respective engine sizes, which was the best available fuel consumption in each size class. Even the body was made from advanced (and secret) materials, which produced a 50% reduction in weight.

Nakagawa concluded his broadcast with the statement that the car industry might become like the television industry where black-and-white TVs became only a miniscule portion of the market when color TVs became widely available. Likewise, conventional cars might lose their appeal after the acceptance of ceramic-engine automobiles. He commented that soon the new cars could be selling like bread, having heard on the news that at one dealership five cars were sold and ten others ordered during a 3-hr period. The World car marketing strategy was similar to a strategy used to market gasoline 20-yr ago when suddenly everyone in Japan woke up to discover that 10,000 service stations had changed ownerships and were displaying new ESSO signs. The M.G. Corporation capitalized on the Japanese marketing techniques by establishing an extensive sales and service network from the beginning and providing a fuel-efficient car which had, in addition, excellent performance characteristics and family car roominess.

The M.G. Corporation had obviously taken advantage of the worldwide depression experienced by the auto industry. Even the Japanese auto dealers had been having trouble, so it was relatively easy for M.G. to take over and establish a distribution network. The Japanese parliment (Diet) had been concerned about their domestic auto industry for some time. Proposed legislation included special low-interest loans guaranteed by the government for the auto industry to develop new technology. Similar legislation was enacted after World War II for the ship-building industry, but scandals resulted. Consequently, the Diet had hesitated to act on this controversial legislation for the auto industry.

In 1981 (four years prior to this introduction of the World car), the President of M.G. had said, "In the near future, we will push the Japanese cars into the Pacific Ocean with our World cars." Mr. Jenkins also said during his interview with Mr. Nakagawa that the Z car, by itself, would not destroy the Japanese auto industry. The problem was that Japan had concentrated on automobiles as exports, but lacked a long-term perspective. The M.G. Company laid the groundwork for the World car by setting up production facilities in 40 countries, including Japan. The M.G. Company had also developed cars in a series to meet specific market needs: first the J car, then the S car, next the T car, and finally the Z car. In response to Mr. Nakagawa's question about whether M.G. had anything in mind to mitigate the adverse effects on the Japanese economy of the Z car, Jenkins replied, "I want to return to the statements of Japanese politians made four years ago. The prime minister and the minister for MITI both said that free trade must be protected and we must not be afraid of trade conflicts." Jenkins added that four years ago the U.S. government did not impose import quotas when Japan had the competitive edge. Now even if the Japanese government wanted to, it could not seriously affect Z car sales by quotas since Z cars would also be made in the Hokkaido (Japan) plant.

During the next day, the television news carried reports that M.G. would begin hiring more employees -- 3000 engineers and other workers for their East Asian operations. This story was followed by reports that Tokai and Nichirin had complained that M.G. was running television ads showing a popular Japanese baseball star driving a World car that passed the Tokai and Nichirin cars on the road. This sort of direct commercial comparison of products was unheard of in Japan.

During the ensuing months, the Japanese auto industry became split in its views concerning how to respond to the American challenge. The smaller companies were either specialized (so that the Z car was not a threat) or entered into agreements with M.G. to produce components for the Z car. Only the big two (Nichirin and Tokai) were adversely affected and they faced a serious threat. Even the Diet could not effectively respond through legislation because the members represented different constituencies. The Liberal Party, for example, was opposed to trade-restricting legislation because small businesses formed an important part of their constituency. However, eventually a controversial, low-interest loan (making $5 billion available to the industry) was enacted, but a scandal resulted. The Big Two auto-makers had bribed several Diet members in order to get the legislation passed.

Opponents of the legislation became concerned that the size of the government involvement could bring about the downfall of the government. The legislation, however, was not a quick-fix for the auto industry and the continuing success of the Z car resulted in serious layoffs at Tokai and Nichirin. Two members of the Diet, from the Social-Democrat party and the Labor-Democrat party were arrested for taking bribes.

The Japanese government sent a mission to ask the U.S. government to restrain excess exports of American small cars for five years. A spokesman for the U.S. said that after the initial 50,000 cars were sold, the Z cars sold in Japan would be produced in Japan. Consequently this issue was out of the reach of the U.S. government.

In the meantime, Tokai and Nichirin, at a secret meeting, proposed that the Japanese auto makers form a cartel to restrict the production of Z car. Kasuaga (presumably Isuzu), the fifth largest company and the one producing the Z car in its Hokkaido plant, refused to go along with the plan. Thus, the nature of the conflict changed from international to domestic.

Newspaper ads, sponsored by the "Domestic Car Protection Group", began appearing and questioned the reliability of using ceramics in car engines. The lighter ceramic cars were consequently presumed to have inherent safety problems. After a fatal traffic accident involving a Z car traveling in excess of 90 mph, a reporter attributed the accident to the light construction of the car. Kasuaga, in retaliation, began its car advertisements illustrating the benefits of ceramics in engines.

With the cartel a failure, Tokai cut production, hoping that the other auto makers would follow suit. This strategy was not successful either. However, world events that would shape the future of Japan's auto industry then took place. North Korea attacked South Korea. The industry saw an opportunity for new production -- military vehicles. The threat of war (and actual small-scale fighting) was also prominent along the Sino-Soviet border, in the Middle East, and in Africa. These small military conflicts became the hope of a desperate industry.

Meanwhile, M.G. continued its successful sales program. The Z car even became a status symbol, so that when prices increased, so did sales. This phenomemon, in turn, prevented the Japanese government from accusing the U.S. of "dumping".

At a top-secret meeting between U.S. and Japanese officials, an agreement was reached for the production of "specialized vehicles." New legislation repealed the post-World War II Japanese constitutional provision that prohibited the export of military vehicles. M.G. and Tokai signed a joint agreement to produce the new military vehicles. However, the general accord was to allow the U.S. to dominate the passenger car market while Japan specialized in military vehicles. Anxiety was reported in some sectors of Japan that the production of military vehicles might result in another global conflict.

References

1. *Ceramic Technology for Advanced Heat Engines, Draft Program Plan*, Oak Ridge National Laboratory Report (March 1983).

2. Teotia, A., Y. Klein, F. Wyant, *Estimating Market Penetration of New Energy Products: A Case Example of New Energy-Efficient Electric Motors*, Proc. Twelfth Annual Modeling and Simulation Conf., University of Pittsburgh, Pittsburgh, Penn. (Apr. 30–May 1, 1981).

3. Teotia, A., C. Lee, A. Kennedy, *Market Penetration of New Energy Systems Estimated by Econometric and Stochastic Methodology*. Proc. Symp. Energy Modeling II, Institute of Gas Technology, Colorado Springs, Colo. (Sept. 1979).

4. Eckstein, O., C. Probyn, *The DRI Scenario Model*, Data Resources U.S. Long-Term Review, Lexington, Mass. (Spring 1982).

5. Sofianou, Z., *The DRI Scenario Model: Structure and Properties*, Data Resources U.S. Long-Term Review, Lexington, Mass. (Spring 1982).

6. Caton, C., *Forecast Summary*, Data Resources U.S. Long Term Review, Lexington, Mass. (Spring 1982).

7. *Industry Series - 1977 Census of Manufacturers*, Bureau of Census, U.S. Department of Commerce, Washington, D.C. (Feb. 1980).

8. U.S. Comptroller, *Learning to Look Ahead: The Need for a National Materials Policy and Planning Process*, General Accounting Office, Washington, D.C. (April 1979).

9. Petkof, B., *Beryllium*, Mineral Commodity Profile, U.S. Bureau of Mines, Washington, D.C. (Aug. 1978).

10. Sibley, S., *Cobalt*, Mineral Commodity Profile 5, U.S. Bureau of Mines, Washington, D.C. (July 1977).

11. *Materials Availability for Automotive Applications*, Society of Automotive Engineers, Warrendale, Penn. (Feb. 1980).

12. Treager, I., *Aircraft Gas Turbine Engine Technology*, McGraw-Hill, New York, N.Y. (1969).

13. Morning, J.L., *Chromium*, Mineral Commodity Profile 1, U.S. Bureau of Mines, Washington, D.C. (May 1977).

14. DeHuff, G., and T.S. Jones, *Manganese*, Mineral Commodity Profile, U.S. Bureau of Mines, Washington, D.C. (July 1979).

15. Corrich, J.D., *Nickel*, Mineral Commodity Profile 4, U.S. Bureau of Mines (July 1977).

16. Jolly, J., *Platinum-Group Metals*, Mineral Commodity Profile 22, U.S. Bureau of Mines, Wasnington, D.C. (Sept. 1978).

17. Jones, T.S., *Tantalum*, Mineral Commodity Profile, U.S. Bureau of Mines, Washington, D.C. (June 1979).

18. Lynd, L., *Titanium*, Mineral Commodity Profile, U.S. Bureau of Mines, Washington, D.C. (Aug. 1978).

19. Stevens Jr., R.F., *Tungsten*, preprint from Bulletin 667, U.S. Bureau of Mines, Washington, D.C. (1975).

20. *Mineral Commodity Summaries-1982*, U.S. Bureau of Mines, Washington, D.C. (1982).

21. Mueller, J.I., *Handicapping the World's Derby for Advanced Ceramics*, American Ceramic Society Bulletin (May 1982).

22. Lenoe, E.M., *Overview of German Structural Ceramics Research*, U.S. Department of Energy, presented at the Energy Materials Coordinating Committee Meeting, Germantown, Md. (Sept. 1982).

23. Gottschall, R.J., *Overview of Japanese "Moonlight" and "Fine Ceramics" Structural Ceramics Research*, U.S. Department of Energy, presented at the Energy Materials Coordinating Committee Meeting, Germantown, Md. (Sept. 1982).

24. *Diesel Rabbit Retains MPG Crown*, Automotive News (Sept. 20, 1982).

25. Mansfield, E., *How Research Pays Off in Productivity*, EPRI Journal, pp. 25-28 (Oct. 1979).

26. Griliches, Z., in *New Developments in Productivity Measurement and Analysis*, J. Kendrick and B. Vaccara, eds., National Bureau of Economic Research, Chicago, pp. 419-454 (1980).

27. Terleckyj, N., *Effects of R&D on the Productivity Growth of Industries, An Exploratory Study*, National Planning Association, Washington, D.C. (1974).

28. Mansfield, E., *Composition of R and D Expenditures: Relationship of Size of Firm, Concentration, and Innovative Output*, Review of Economics and Statistics, LX 111-4, pp. 610-612 (Nov. 1981).

29. Mansfield, E., *Basic Research and Productivity Increase in Manufacturing*, American Economic Review, 70:5 (Dec. 1980).

30. *Japan's High-Tech Challenge*, Newsweek, (Aug. 9, 1982).

31. Data Resources U.S. Long-term Review, Lexington, Mass. (Fall 1982).

Other Noyes Publications

FRACTURE IN CERAMIC MATERIALS
Toughening Mechanisms, Machining Damage, Shock

Edited by
A.G. Evans
Department of Materials Science and Mineral Engineering
University of California, Berkeley

This book presents recent studies on the mechanisms of fracture in ceramic materials—the effects of toughening, machining, and shock. Research on toughening mechanisms, machining and surface damage, thermal shock and general aspects of fracture in ceramic materials is described. Quantitative models of the various fracture processes have been developed. Special emphasis has been placed on the toughening that occurs in the presence of microcracks.

During the last decade, research on the fracture of monolithic single phase and multiphase ceramic polycrystals has attained a maturity which now permits many fracture phenomena to be quantitatively described. Specifically, the predominant fracture initiating flaws have been identified and the fundamental mechanics and statistics related to their fracture severity have been determined. In addition, the crack growth resistance exhibited by common ceramic microstructures can now be expressed in quantitative terms, through the development of micromechanics models of transformation toughening, microcrack toughening, and deflection toughening.

As a result, the next research frontier in the field of advanced monolithic ceramics undoubtedly resides in studies of the processing of optimum microstructures. Progress in this area is summarized in this book. The four main subject areas of the book are toughness/microstructure interactions, machining damage, thermal fracture and reliability, and impact damage.

A condensed table of contents listing **part and chapter titles** is given below.

I. TOUGHNESS/MICROSTRUCTURE INTERACTIONS

1. **TOUGHENING MECHANISMS IN ZIRCONIA ALLOYS**
2. **THE MECHANICAL BEHAVIOR OF ALUMINA: A MODEL ANISOTROPIC BRITTLE SOLID**
3. **OBSERVATIONS OF INTERGRANULAR, CRACK DEFLECTION TOUGHENING MECHANISMS IN SILICON CARBIDE**
4. **ON THE CRACK GROWTH RESISTANCE OF MICROCRACKING BRITTLE MATERIALS**
5. **MICROSTRUCTURAL RESIDUAL STRESSES**
6. **SPONTANEOUS MICROFRACTURE IN MICROSTRUCTURAL RESIDUAL STRESSES**
7. **INDUCED MICROCRACKING: EFFECTS OF APPLIED STRESS**

II. MACHINING DAMAGE

8. **FAILURE FROM SURFACE FLAWS**
9. **SURFACE FLAWS IN GLASS**
10. **MECHANISMS OF FAILURE FROM SURFACE FLAWS IN MIXED MODE LOADING**
11. **GEOMETRICAL EFFECTS IN ELASTIC/PLASTIC INDENTATION**
12. **RESIDUAL STRESSES IN MACHINED CERAMIC SURFACES**
13. **FATIGUE STRENGTH OF GLASS: A CONTROLLED FLAW STUDY**

III. THERMAL FRACTURE AND RELIABILITY

14. **THE THERMAL FRACTURE OF ALUMINA**
15. **ASPECTS OF THE RELIABILITY OF CERAMICS FOR ENGINE APPLICATIONS**

IV. IMPACT DAMAGE

16. **LENGTH OF MAXIMAL IMPACT DAMAGE CRACKS AS A FUNCTION OF IMPACT VELOCITY**

ISBN 0-8155-1005-5 (1984)   420 pages

HORACE BARKS
REFERENCE LIBRARY

STOKE-ON-TRENT